Underwater embankments on soft soil: a case history

BALKEMA – Proceedings and Monographs
in Engineering, Water and Earth Sciences

Underwater embankments on soft soil: a case history

W.F. Van Impe
R.D. Verástegui Flores

University of Ghent, Laboratory of Geotechnics,
Zwijnaarde, Belgium

Taylor & Francis
Taylor & Francis Group

LONDON / LEIDEN / NEW YORK / PHILADELPHIA / SINGAPORE

Taylor & Francis is an imprint of the Taylor & Francis Group, an informa business

© 2007 Taylor & Francis Group, London, UK

Reproduced as CRC by Charon Tec Ltd (A Macmillan Company), Chennai, India
Printed and bound in Hungary by Uniprint International bv (a member of the Giethoorn Media-group) Székesfehérvár.

All rights reserved. No part of this publication or the information contained herein may be reproduced, stored in a retrieval system, or transmitted in any form or by any means, electronic, mechanical, by photocopying, recording or otherwise, without written prior permission from the publishers.

Although all care is taken to ensure integrity and the quality of this publication and the information herein, no responsibility is assumed by the publishers nor the author for any damage to the property or persons as a result of operation or use of this publication and/or the information contained herein.

Published by: Taylor & Francis/Balkema
P.O. Box 447, 2300 AK Leiden, The Netherlands
e-mail: Pub.NL@tandf.co.uk
www.balkema.nl, www.taylorandfrancis.co.uk, www.crcpress.com

Library of Congress Cataloging-in-Publication Data

Van Impe, W.F.
   Underwater embankments on soft soil : a case history / W.F. Van Impe,
R.D. Verástegui Flores.
   p. cm.
   Includes bibliographical references.
   ISBN 978-0-415-42603-9 (hardcover : alk. paper)
1. Embankments. 2. Soil stabilization. 3. Underwater construction. I. Verástegui Flores, R. D. II. Title.

TA760.V3576 2007
627′.58–dc22
                                        2007001786

ISBN: 978-0-415-42603-9(Hbk)
ISBN: 978-0-203-94629-9(eBook)

Contents

About the Authors IX

Preface XI

    List of symbols 1

1 Introduction 1

2 Construction on soft soil 3
    2.1 Clay foundation behavior 3
    2.2 Staged construction 5
        2.2.1 Methodology for analysis 7
        2.2.2 Consolidation analysis 7

3 Discussion on undrained shear strength of soft soils 11
    3.1 Preamble 11
    3.2 Shear behavior of normally consolidated clay 11
    3.3 Normalized behavior 13
    3.4 Shear behavior of overconsolidated clay 14
    3.5 Laboratory testing techniques 16
        3.5.1 Overview 16
        3.5.2 SHANSEP 18
        3.5.3 Discussion 19
    3.6 Undrained strength anisotropy 20
    3.7 Final recommendations 23

4 Discussion on slope stability evaluation 25
    4.1 Preamble 25
    4.2 Causes of slope instability 26
    4.3 Stability conditions for analysis 28
    4.4 Stability analysis procedures 30

VI Contents

    4.4.1 Limit equilibrium methods 30
    4.4.2 Strength reduction methods 31
    4.4.3 Limit equilibrium versus strength reduction methods. . . 33
  4.5 Failure mechanisms for highly sensitive clays 35
    4.5.1 Flake type sliding of quick clay. 37
    4.5.2 Analysis taking pore water pressures into account 40
    4.5.3 Mechanism of sliding in quick clay masses 44
    4.5.4 Conclusions. 49
  4.6 Risk of liquefaction 49
  4.7 Slope stability analysis of the Doeldok embankment. 51

5 Evaluation of consolidation 57
  5.1 One-dimensional consolidation theory 57
  5.2 Infinitesimal strain theory 58
    5.2.1 Numerical solution 60
    5.2.2 Applications of SSCON-FD 62
  5.3 Finite strain theory 64
  5.4 Infinitesimal strain versus finite strain theory 66
  5.5 Consolidation at the Doeldok site. 66
  5.6 Conclusions. 70

6 Geotechnical characterization of the site 71
  6.1 Overview 71
  6.2 Soil profile and characterization 71
  6.3 Selection of parameters for design 72
    6.3.1 Dredged material. 73
    6.3.2 Boom clay. 78
    6.3.3 Sand. 79
    6.3.4 Summary of soil properties 80

7 Design of underwater embankment of soft soil 81
  7.1 Overview 81
  7.2 Geometry of the embankment 82
  7.3 Stability analysis 83
    7.3.1 Undrained analysis 83
    7.3.2 Drained stability analysis. 86
  7.4 Settlements 87
    7.4.1 From Constitutive relationships 87
    7.4.2 From finite element program 88
  7.5 Conclusions. 88

8 Ground improvement by deep mixing 91
  8.1 Introduction 91
  8.2 Deep mixing applications 92
  8.3 Mechanism of stabilization. 93

	8.3.1	Stabilization with lime 94
	8.3.2	Stabilization with cement-like binders 94
8.4	Methods of installation 96	
8.5	Belgian experience on on-land deep mixing 97	
	8.5.1	Properties of untreated soils on land 98
	8.5.2	Binders employed on land 100
	8.5.3	Lime-cement stabilization in the laboratory 100
	8.5.4	Lime-cement stabilization in-situ 102
	8.5.5	Remarks on the experience of dry deep mixing on land . 107
8.6	Deep mixing assessment on the underwater site 109	
	8.6.1	Properties of the artificially cemented soil in the laboratory 110
	8.6.2	Properties of the cemented soil in the field 116
	8.6.3	Laboratory versus in-situ behavior 119

9 Construction and monitoring of embankment 123
9.1 Introduction 123
9.2 Construction 123
9.3 Quality control of the embankment sand 124
9.4 Instrumentation and monitoring 128
    9.4.1 Excess pore water pressure 129
    9.4.2 Settlements 133
9.5 Conclusions 134

References 135

Subject Index 139

About the Authors

William Frans Van Impe

Professor Van Impe was appointed Professor in the Faculty of Applied Sciences of Ghent University in 1982 and has established a Laboratory of Soil Mechanics which enjoys both a national and international reputation. Since 1988, he has been a visiting professor at the Katholieke Universiteit Leuven and is member of the Board of Directors of the Royal Academy of Overseas Sciences and board member of the European DFI group. From 1994-1998 he was vice-president of the ISSMGE, Europe and in 2001 became President of the ISSMGE for a four-year term. His research has been rewarded with several international prizes and awards. He and his team have authored three books and over 200 papers in journals and proceedings. Professor Van Impe has acted as a geotechnical consultant to numerous overseas projects and was the Geotechnics Committee member of the project evaluation of the Messina Strait Crossing - the largest suspension bridge in the world.

Ramiro Daniel Verástegui Flores

R. D. Verástegui Flores was born in Bolivia in 1976. He graduated as a Civil Engineer in 1999 at San Simon University (UMSS) in Bolivia where he was awarded a prize in recognition of his research on soil dynamics by the Faculty of Technology and Science at UMSS. In 2002 he completed a Post-graduate Degree on Geotechnical Engineering at Ghent University in Belgium under the guidance of Prof. W. F. Van Impe. Since then, he has been involved in research projects dealing with soft soils engineering in near shore and offshore conditions. Currently he is an Academic Assistant and PhD student at Ghent University. He is author and co-author of a number of papers in international conference proceedings and journals. His interests focus on stress-strain behaviour of cemented soils, ground improvement techniques, large strain consolidation of soft soils and deep foundations.

Preface

Ground improvement is probably the oldest but, from a technical point of view, still the most intriguing technique of all common execution methods in foundation engineering. Indeed, ground improvement was already in use more than 3000 years ago. In recent decades, the modern methods of ground improvement are making use of explosives, impact energy, thermal treatment of the soil, vacuum consolidation, vibratory compaction technologies, stabilization and solidification of soft soils, as well as combined systems of ingenious grouting systems, deep mixing techniques, etc.

Throughout the world, deep mixing techniques today are of utmost importance in dealing with more and more demanding foundation problems. This tendency has been noticed in Belgium already at a very early stage; with inventive new developments of soft soil deep mixing technologies and various advanced high pressure mixing methods. Some initial experiences onshore and offshore have proved already some years ago that successful solutions can be attained.

The present work illustrates a challenging example of design and construction of a quite important large underwater embankment on very soft soil. Throughout the design staged construction and ground improvement by deep mixing combined with geotextile reinforcement were proposed to assure the safety as well as allowable deformations of the construction.

The outcome of monitoring excess pore water pressures and displacements during the construction shows that when taking account of key aspects of advanced soil stress-strain behavior, it is possible to appropriately model such complex problem and even to make rather simple attempts to reach some of the "type A" foundation behavior predictions.

We may expect this work to be recognized as a valuable reference case history for the geotechnical engineer, both from the academic as well as from the practitioner's point of view, in order to contribute to the art of building on soft soils.

Our acknowledgments go to the contractors DEME and J. DE NUL, to the teams of geotechnical experts of Dredging International NV, HydroSoil

Services, to the Geotechnical Division and Maritime Access Division of the Flemish Ministry, all actively contributing to the satisfactory result of this uncommon foundation engineering problem. The expert group following closely the design and construction of the embankment is listed here:

J. Van Mieghem	Flemish Ministry, Department of Maritime Access
H. De Preter	Flemish Ministry, Department of Maritime Access
J. Van den Broecke	Flemish Ministry, Department of Maritime Access
C. Boone	Technum, Belgium
H. Cecat	Technum, Belgium
M. Van den Broeck	Dredging International
R. Aelvoet	Combinatie Kallo
P. Mengé	Combinatie Kallo
S. Vandycke	Combinatie Kallo
F. Verhees	Combinatie Kallo
R. Lheureux	Combinatie Kallo
P. De Schrijver	Flemish Ministry, Department of Geotechnics
A. Baertsoen	Flemish Ministry, Department of Geotechnics
R. Simons	SECO, Belgium
R. Dedeyne	SECO, Belgium

In each of the various chapters of the book, the discussion contributing expert group members have been mentioned.

Ghent
February 2007

W.F. Van Impe
R.D. Verástegui F.

List of symbols

ϕ'	[°]	shear angle
σ_1, σ_3	[kPa]	principal stresses
σ'_p	[kPa]	preconsolidation pressure
σ'_v	[kPa]	vertical effective stress
σ'_{v0}	[kPa]	initial vertical effective stress
CRR		cyclic resistance ratio
CSR		cyclic stress ratio
c'	[kPa]	effective cohesion
c_u	[kPa]	undrained shear strength
c_v	[m^2/s]	coefficient of consolidation
e		void ratio
E'	[MPa]	drained Young's modulus
E_u	[MPa]	undrained Young's modulus
f_s	[MPa]	cone sleeve friction
k	[m/s]	hydraulic conductivity
K_0		horizontal stress coefficient at rest
m_v	[1/kPa]	coefficient of volume compressibility
p'		effective mean stress
q'		deviatoric stress
q_c	[MPa]	cone penetration pressure
q_u	[kPa, MPa]	unconfined compressive strength
R		shear strength reduction factor
r_k		anisotropic hydraulic conductivity ratio
S		undrained strength ratio
u	[kPa]	pore water pressure

1

Introduction

W.F. Van Impe & R.D. Verástegui Flores
Laboratory of Geotechnics, Ghent University, Belgium

J. Van Mieghem
Maritime Access Division, Ministry of Flanders, Belgium

As in many harbor area all around the world, there is an increasing need of reclaimed land for storing excavated soil in the harbor of Antwerp in Belgium, mainly because of the construction of new docks. This fact has encouraged the partial filling up of a dock with a partially submerged embankment as a retaining structure. Figure 1.1 illustrates the location of the embankment in the dock (Doel dock) at the harbor of Antwerp. The design and ongoing construction of a partially submerged 27-m high sand embankment to be founded in this case on about 8m of very soft soil (not removable because of geoenvironmental considerations) created an uncommon challenge.

The soil profile at the site consists of an upper soft layer overlying a thin tertiary sand and a very thick tertiary overconsolidated clay layer. Chapter 6 shows in more detail the geotechnical characterization of the site. The soft material is the result of years of sedimentation and self weight consolidation of dredged material from the harbor waterways. Out of preliminary field and lab

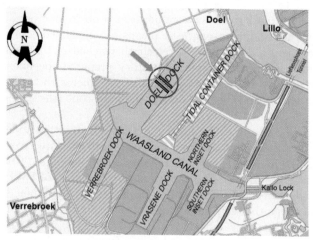

Figure 1.1. Location on the embankment in the Doel dock

testing it was found that this dredged material still remained in a NC state with an initial shear strength of about $c_u = 3\,\text{kPa}$. The consolidation behavior of the soft material was studied with more precision later on through CRS, oedometer and hydraulic conductivity tests.

Throughout the design phase (see Chapter 7) numerous preliminary design options were worked out aiming at optimizing the dam design. Given the soft consistency of the dredged material, it became clear that some kind of foundation layer reinforcement was needed. Therefore ground improvement by a novel deep mixing technology (SSI) was proposed, mainly applied at the toes of the embankment. These improved zones are meant to provide extra strength and to serve as a means of confinement of the soft soil in the middle part of the embankment foundation layer.

The analysis showed, as expected, that the construction phase was actually the most critical stage for the embankment stability. Unavoidably, a staged construction was implemented. Staged construction is a technique that uses controlled rates of loading to enable soil strength to develop via consolidation in order to increase the embankment stability. Consequently, an accurate evaluation of the consolidation progress (taking account of large deformations) had to be the key issue for this foundation problem on very soft soil.

The soft material was improved by deep mixing columns introducing the SSI technique. This deep mixing technique (SSI) makes use of a combination of highly pressurized cement slurry jetted through a set of nozzles along the front side of a rotating arm on the one hand and a series of low pressure cement slurry nozzles on the back side of the rotating arm on the other hand. The rotation of the arm is combined with a continuous uplift movement; the cement mixing in this respect is uniformly distributed along the diameter of the column.

An extensive laboratory research program was set up to study the behavior of the improved material (see Chapter 8). A number of cement type were tried out and strength measurements were performed up to long periods of curing. Control of actual column strength by means of unconfined compression tests on core specimens showed that the strength reached in the field was higher than the strength expected from laboratory prepared specimens. This fact was also thoroughly explained by looking at the microstructure of specimens from the laboratory and the field under the microscope (the scanning electron microscope).

The behavior of the embankment, still in construction today, has been assessed by means of elaborated instrumentation to measure excess pore water pressure (PWP) and settlements of the foundation layer (see Chapter 9). As expected, the measurements do show a slow dissipation of PWP; the excess PWP measured within the SSI-column improved zone is by far smaller than PWP within the untreated zone because of the stiffness of the SSI treated soil columns. As for the vertical displacements and PWP observed so far, they are in good agreement with the expected values from the design.

2

Construction on soft soil

W.F. Van Impe & R.D. Verástegui Flores
Laboratory of Geotechnics, Ghent University, Belgium

2.1 Clay foundation behavior

Construction on soft soils remains a troublesome foundation engineering challenge as very often a significant load is imposed by the new structure and the available shear strength of the soft clay is low.

Traditionally, the analysis of such problem is made in two steps that consist of:

- Slope stability analysis during construction under fully undrained conditions.
- Slope stability analysis long after the end of construction under fully drained conditions.

The validity of this approach was shown not always satisfactory as reported already by Leroueil et al. (1990) after they compared actual measurements and predictions assuming fully undrained conditions. Such observations, for different soil conditions, can be explained looking at the stress path of an element under the centerline of e.g. an embankment load.

Figure 2.1 illustrates the case of an embankment on slightly overconsolidated (OC) clay. The most likely stress path evaluated from many observations is depicted by the path ABC. The stress state of the soil starts at A and reaches the yield surface in a zone (B) where the vertical effective stress σ'_v is approximately equal to the preconsolidation of the clay σ'_p. Throughout path AB, the coefficient of consolidation is usually high; then, a relatively fast dissipation of excess pore water pressure can be expected. However, once the soil reaches the yield surface it becomes normally consolidated (NC) and a rather significant drop of the coefficient of consolidation is usually observed. Therefore, only the BC path may be considered essentially as an undrained (slow dissipation of PWP).

Similarly, the behavior of a highly OC clay foundation is illustrated in figure 2.2. The initial state of stress is closer to an isotropic state (see point A).

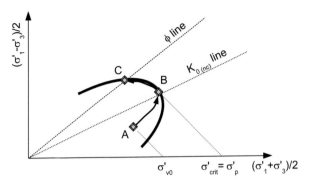

Figure 2.1. Behavior of a slightly OC clay foundation

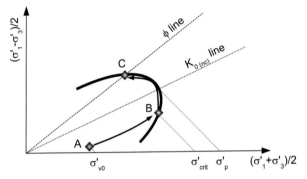

Figure 2.2. Behavior of a highly OC clay foundation

Observations show that during "undrained" loading, overconsolidated clays generate lower pore pressures than soft clays. Their high stiffness implies a high coefficient of consolidation. Unlike slightly OC clays, typical stress paths of highly OC clays reach the yielding surface (B) at a value of vertical effective stress lower than the preconsolidation pressure. Any further loading follows a path along the yield surface (BC) where slow dissipation of excess pore water pressure can actually occur.

As a result, the assumption of fully undrained conditions becomes valid only for normally consolidated clays (see Fig. 2.3) since their initial stress state falls already in the yielding surface. In fact, many experiences show trends of excess pore water pressure measurements such as those illustrated in figure 2.4. In NC clays, the excess pore water pressure increases at about the same rate as the total applied stress (undrained conditions), however, in OC clays the pore pressures generated at the start of the construction are low and then rise roughly at the same rate as the total applied vertical stress once the embankment has reached a critical loading (related to the yielding of the soil).

In the specific case of the design of the underwater embankment, it was observed that the foundation soil is almost fully normally consolidated, then assuming fully undrained conditions for stability analysis is not mistaken.

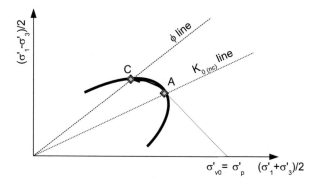

Figure 2.3. Behavior of a NC clay foundation

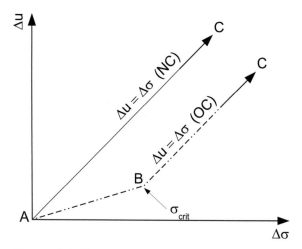

Figure 2.4. Excess pore water pressure mobilization

Sometimes, as in the present case, it is necessary to go beyond the capacity of the clay. Here, the total load imposed by the embankment is greater than the bearing capacity of the foundation soil. To tackle this problem the engineer is left with two options, either to choose for some kind of ground improvement technique to increase the soft soil strength or to work with the soil allowing it to drain and increase its own strength by running the construction at a controlled rate. This second option is called staged construction and it is discussed in detail in the next section.

2.2 Staged construction

Staged construction is a technique employed in soft soil construction, where the imposed loading is sufficiently large to stress the cohesive foundation soils beyond their preconsolidation pressure and close to failure. Examples include

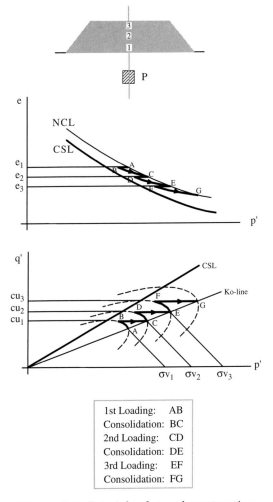

Figure 2.5. Principle of staged construction

embankments for transportation facilities, flood-control levees, tailing dams, landfills and offshore structures.

Since these projects generate positive excess pore water pressure within the foundation soil, the most critical condition occurs during construction; therefore, undrained stability analysis are commonly implemented. As previously discussed, the assumption of fully undrained conditions ($\Delta\sigma = \Delta u$) for stability analysis is appropriate only for saturated normally consolidated clays.

In order to have a better understanding of the behavior of the foundation soil subjected to a load, figure 2.5 outlines the effective stress path of a normally consolidated element located under the central line of an embankment constructed in stages.

We can see that the total equivalent load applied (represented by point G) is greater than the initial bearing capacity of the soft soil foundation; if the construction would be carried out in a single step, failure would indeed occur. Then, a staged construction has to be implemented to ensure safety during construction.

Three stages have been outlined, the effective stress path for the first loading stage is AB. Loading is halted before failure occurs. The initially available undrained strength is c_{u1}. If full consolidation is allowed, the excess pore water pressure dissipates and the effective stress path goes from B to C. Consolidation leads to a reduction of the void ratio as density increases.

Note that for the second loading stage (path CD) an increased undrained strength c_{u2} becomes available now. Again, at the end of the second stage (point D) consolidation is allowed (path DE) and finally, the third stage EF can be safely constructed. However, the drawback of the staged construction method relates to the more extended period for construction required. When construction time is not an issue, the implementation of the staged construction principle in a project is a cost effective solution.

2.2.1 Methodology for analysis

The design of staged construction projects usually entails the steps summarized in figure 2.6. First it is necessary to evaluate stability for the first stage loading assuming no drainage during construction, based on the initial mechanical parameters of the soil. Then, stability calculations during subsequent construction stages can be made taking into account that the combination of the previously applied loads and either partial or full consolidation will change the initial stress history and will increase the available strength of the foundation soil.

In figure 2.6, the initial state variables refer to the soil profile and pre-consolidation pressure. These, together with laboratory testing results (to measure undrained strength) supply the basic information.

The first stability analysis makes use of initial soil data to compute the factor of safety of stage 1. Subsequent stability evaluations require knowledge of consolidation degree in order to predict a new undrained strength.

2.2.2 Consolidation analysis

Clearly, staged construction design requires an accurate consolidation analysis to predict rates of pore water dissipation during construction. These predictions often have a strong impact on project feasibility, schedule and costs during design.

In the following paragraphs some problems regarding consolidation evaluation are identified.

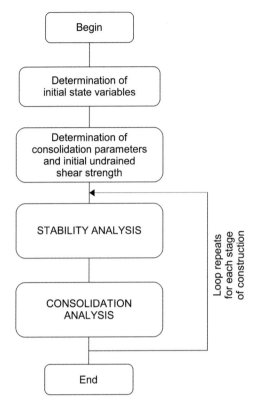

Figure 2.6. Methodology for staged construction analysis

2.2.2.1 One-dimensional consolidation

Although practice often relies on the conventional 1D small strain consolidation theory after Terzaghi (1925), the assumption of constant values of coefficient of consolidation (c_v) and compressibility (m_v) may give poor estimates (non conservative) of pore water pressure dissipation during construction, especially when dealing with very soft soils in which parameters may significantly change with strain. Problems may also arise when dealing with moderately overconsolidated deposits (which suffer large changes in both c_v and m_v near the overconsolidation pressure p') and layered clay deposits having different c_v and m_v values.

Such situations do require a modeling technique that allows the use of non linear functions to describe the consolidation parameters, changes of load with time and less restrictive boundary conditions. Chapter 5 examines this topic in more detail.

2.2.2.2 Accelerating consolidation with drains

The primary purpose of vertical drains, often implemented together with staged construction, is to speed up the consolidation process by allowing horizontal drainage and shortening the drainage path as well. In addition, most natural deposits are anisotropic with respect to flow properties, the horizontal hydraulic conductivity being typically higher than the vertical one.

Installation of vertical drains to accelerate the rate of consolidation introduces two additional problems regarding reliable predictions of pore pressure during staged construction: evaluation of the anisotropic permeability ratio $r_k = k_h/k_v$, required to estimate the coefficient of consolidation for horizontal flow; and assessment of the likely effects of soil disturbance caused by drain installation (smear effect), depending on the particular characteristic of soil and drain to install. Analytical solutions have been studied by many researchers (Barron, 1948; Hansbo, 1979; Van Impe, 1989; Lancellotta, 1995). However, all of them assume basically constant consolidation parameters in time.

3

Discussion on undrained shear strength of soft soils

W.F. Van Impe & R.D. Verástegui Flores
Laboratory of Geotechnics, Ghent University, Belgium

3.1 Preamble

The execution of a staged construction design on soft normally consolidated soil implies the assessment of the stability in undrained conditions. Therefore, the initial in-situ undrained shear strength (c_u) of the cohesive foundation soil and the subsequent changes of c_u during construction have to be estimated. In this chapter, some issues linked to the shear strength of soils and progressive gain with consolidation are brought to discussion.

3.2 Shear behavior of normally consolidated clay

The shear strength of normally consolidated clays in both drained and undrained conditions is traditionally represented by Mohr-Coulomb envelopes, characterized by a friction angle ϕ' and an effective cohesion c' which is usually negligible. Data from a large number of clays from various origins show that the friction angle is a function of the plasticity of the clay and is independent of the rate of strain (Fig. 3.1).

Normally consolidated clays behave as plastic materials, consequently, their stress strain behavior is governed by plastic flow rules; therefore, the definition and use of an "elastic" modulus of deformation is not justified (Tavenas, 1987). Yet, the stress-strain behavior of normally consolidated clay is often expressed in terms of undrained modulus (e.g. E_u).

According to the soil yielding concept, a normally consolidated sample sheared under undrained conditions follows an effective stress path along the yielding surface. If the sample is one dimensionally consolidated, as occurs in nature, the yielding surface is as that shown in figure 3.2.

Roscoe et al. (1958) have shown, based on isotropically consolidated remolded clay, that the yielding surface may be described as an elliptical curve centered on the isotropic axis (Cam-Clay model). However tests on natural

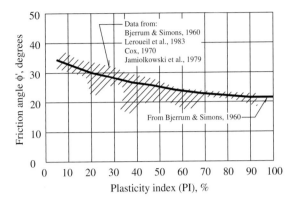

Figure 3.1. Variation of the effective friction of NC clay with the Plasticity Index (Mesri and Abel-Ghaffar, 1993)

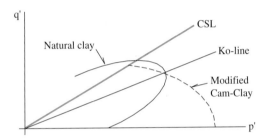

Figure 3.2. Typical yielding surface of natural clay

clay, which may be assumed to be K_0-consolidated, show a yield surface centered on the K_0-line (Fig. 3.2). The shape of the yield surface is influenced by K_0 which in turn is determined by ϕ', closely related to the plasticity index.

Figure 3.3 shows the effective stress path of a set of three samples of clay, K_0-consolidated at different vertical stresses, subjected to undrained compression. In the figure, $p' = (\sigma_1' + \sigma_3')/2$ and $q' = (\sigma_1' - \sigma_3')/2$. By geometrical means we can identify the initial vertical effective stress on each sample and at the same time we can identify the undrained strength obtained after shearing. It can be observed that for a normally consolidated clay, the ratio c_u/σ_v' is a constant, i.e.:

$$S = \left[\frac{c_u}{\sigma_v'}\right]_{(1)} = \left[\frac{c_u}{\sigma_v'}\right]_{(2)} = \left[\frac{c_u}{\sigma_v'}\right]_{(3)} \tag{3.1}$$

This important founding led to a better understanding of the clay behavior subjected to undrained shear. Moreover, it has led to one of the most important concepts in soil mechanics, normalized behavior, which is described in the next section.

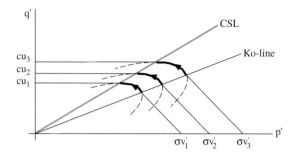

Figure 3.3. Effective stress path in undrained triaxial compression on NC samples

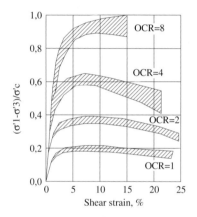

Figure 3.4. Normalised stress-strain behaviour in triaxial testing (from Ladd and Foot, 1974)

The undrained strength ratio $S = c_u/\sigma'_v$ obtained from triaxial compression tests does not vary significantly with the plasticity index and is typically within a narrow range. A typical value of S is 0.3 (Tavenas, 1987).

3.3 Normalized behavior

The work carried out by Roscoe et al. (1958), Henkel (1960) and Ladd and Foot (1974), mostly on reconstituted and destructured clays has shown that undrained tests on samples having the same overconsolidation ratio but different maximum past pressures σ'_p present similar stress-strain and pore pressure-strain characteristics when normalized with respect to the consolidation stress σ'_c or the maximum past pressure σ'_p. Figure 3.4 presents data from Ladd and Foot (1974).

These observations gave birth to what Ladd and Foot (1974) called later the Normalized Soil Parameter (NSP) concept. This concept is very powerful since it applies to all kind of tests: triaxial, plane strain, direct shear, etc., in undrained as well as in drained conditions. The NSP is one of the most fundamental concepts in soil mechanics.

In practice, normalized behavior is not as perfect as that shown in figure 3.4. There is usually some divergence in the normalized plots obtained for different consolidation stresses and also some due to heterogeneity in the soil deposit. Inevitable minor variations in the procedure from one test to another can also cause divergence.

Anyhow, normalized behavior has been found to apply to a wide range of cohesive destructured soils. Still, it is recommendable to check if it is valid for a more specific case.

Quick clays and highly structured clays will not show a good normalized behaviour because their structure is usually altered when subjected to reconsolidation beyond the current stress.

3.4 Shear behavior of overconsolidated clay

When dealing with overconsolidated clays it should be observed that during unloading the soil has lower water content at the same effective stress than a normally consolidated material. As a consequence the undrained strength ratio S, previously defined, should increase with the overconsolidation ratio (OCR). From data collected by Ladd and Foot (1974) and as shown in figure 3.5, it appears that the undrained shear strength of destructured clays varies with the overconsolidation according to:

$$\left[\frac{c_u}{\sigma'_v}\right]_{OC} = \left[\frac{c_u}{\sigma'_v}\right]_{NC} \cdot OCR^m = S \cdot OCR^m \qquad (3.2)$$

Ladd and Foot (1974) reported m values approximately equal to 0.8. Jamiolkowski et al. (1985) indicated that the preceding equation could also be a good approximation for intact natural clays. Equation 3.2 is also supported by the critical state theory as shown below.

The critical state line can be defined in the space $p' : q'$ and $v : p'$ as follows (Atkinson and Bransby, 1978):

$$q' = M \cdot p' \qquad (3.3)$$

$$v = \Gamma - \lambda \cdot \ln p' \qquad (3.4)$$

where, v is the specific volume, Γ is the specific volume at $p' = 1\,\text{kPa}$ and λ is the slope of the normal consolidation line. M, Γ, λ and κ, defined in figure 3.6, are soil constants.

Let's consider two specimens A and B as shown in figure 3.6. Specimen A is NC while specimen B has a $OCR = p'_A/p'_B$. The failure states on the critical state line are indicated by points C and D respectively. The undrained strength

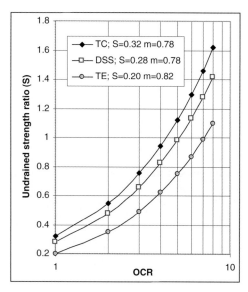

Figure 3.5. Undrained strength ratio ($S = c_u/\sigma'_v$) from CK_0U tests (Jamiolkowski et al., 1985)

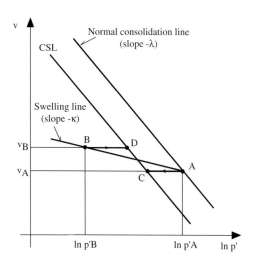

Figure 3.6. Undrained shear compression of overconsolidated sample (Atkinson and Bransby, 1978)

$(c_u)_A$ of sample A, which fails on the critical state line with an effective mean normal stress p'_C and specific volume v_A, is:

$$(c_u)_A = M \cdot p'_C = M \cdot e^{\frac{\Gamma - v_A}{\lambda}} \tag{3.5}$$

The swelling line AB passes through point A and has the following equation:

$$v = v_A + \kappa \cdot \ln \frac{p'_A}{p'} \tag{3.6}$$

Now, if sample B is allowed to swell to a specific volume v_B, by analogy with equation 3.5, the undrained strength $(c_u)_B$ will be:

$$(c_u)_B = M \cdot e^{\frac{\Gamma - v_B}{\lambda}} \tag{3.7}$$

Equation 3.6 can be substituted in equation 3.7 putting $v = v_B$, it gives:

$$(c_u)_B = M \cdot e^{\frac{\Gamma - v_A}{\lambda} - \frac{\kappa}{\lambda} \ln \frac{p'_A}{p'_B}} \tag{3.8}$$

$$(c_u)_B = M \cdot e^{\frac{\Gamma - v_A}{\lambda}} \left(\frac{p'_B}{p'_A}\right)^{\frac{\kappa}{\lambda}} = (c_u)_A \left(\frac{p'_B}{p'_A}\right)^{\frac{\kappa}{\lambda}} = (c_u)_A \, OCR^{\frac{-\kappa}{\lambda}} \tag{3.9}$$

If we divide the previous equation by p'_B, and keeping in mind that $OCR = p'_A/p'_B$, we obtain:

$$\frac{(c_u)_B}{p'_B} = \frac{(c_u)_A}{p'_B} OCR^{-\frac{\kappa}{\lambda}} = \frac{(c_u)_A}{p'_A} OCR^{1-\frac{\kappa}{\lambda}} \tag{3.10}$$

$$\frac{(c_u)_B}{p'_B} = \frac{(c_u)_A}{p'_A} OCR^{\Lambda} \quad \text{with} \quad \Lambda = \frac{\lambda - \kappa}{\lambda} \tag{3.11}$$

We can see that equation 3.11 shows a close relationship with equation 3.2 since we are relating normalized parameters of OC to NC clay. The plastic volumetric strain ratio Λ, as named by Schofield and Wroth (1968), is determined from the slopes of the compression and swelling lines. Its value is limited to be between 0 and 1, and is typically about 0.8.

3.5 Laboratory testing techniques

3.5.1 Overview

The main laboratory testing technique developed for the use with the NSP concept involves consolidation to stresses in excess of those in-situ in order to overcome sample disturbance effects and to control the OCR.

The simplified effect of sample disturbance is shown in the idealized void ratio vs. log effective stress plot illustrated in figure 3.7. The virgin compression curve is typically a unique relationship for a specific clay, time of consolidation and type of consolidation stress system. If a sample becomes overconsolidated, its effective stress is reduced and it swells, typically, following a relationship such as line a in the figure. With reconsolidation, the relationship will follow line b back to the virgin compression line.

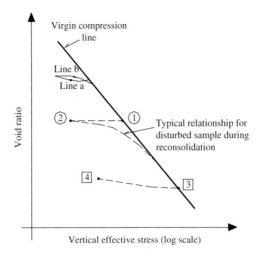

Figure 3.7. Idealised plot showing the effect of sample disturbance (Ladd and Foot, 1974)

Since the changes in void ratio associated with soil swelling are much smaller than those associated with virgin compression, overconsolidated soils always plot below the virgin compression line.

An "undisturbed" sample will typically suffer a decrease in effective stresses during sampling even though the water content may be kept virtually constant. Thus, an in-situ normally consolidated sample at point 1 (in Fig. 3.7) might plot at point 2 after sampling and be similar to an overconsolidated sample. With reconsolidation it will follow some path back to the virgin compression line (dashed line in Fig. 3.7). It follows that a test performed at conditions corresponding to any point on this line prior to its reaching the virgin compression line (e.g. a sample reconsolidated to the in-situ stress) has an uncertain OCR.

On the other hand, a sample that has been consolidated back to the virgin compression line has a clearly known OCR = 1. This sample will give NSP values which, assuming the concept holds for the soil, are equally applicable to all normally consolidated samples. If NSP values for overconsolidated samples are required, these can be obtained at known OCR values by consolidating the samples back to the virgin compression line and then reducing the effective stress to the required OCR. This is shown in figure 3.7 as consolidation from point 2 to point 3, followed by unloading to point 4 to give a sample of known OCR.

Thus the testing procedure to yield NSP values requires that the samples be consolidated back to the virgin compression line before testing. Consolidation to stress levels greater than 1.5 to 2 times the in-situ stress are recommended by Ladd and Foot (1974). Moreover, they provide a procedure:

1. Consolidate samples to approximately 1.5 times, 2.5 times and 4 times the in-situ vertical stress and measure c_u/σ'_v. A clay exhibiting normalized

behavior will yield a constant value of c_u/σ'_v, at least at the two higher stresses. If c_u/σ'_v varies consistently with stress, the NSP concept does not apply to the clay.

2. To obtain c_u/σ'_v vs. OCR, use the minimum value of σ'_v giving normalized behavior as the laboratory preconsolidation pressure and perform tests at OCR values of 2 ± 0.5, 4 ± 1 and 6 ± 2. Compare the results to those plotted in figure 3.5 to check their reliability. The data point should form a smooth concave upward curve

It follows that use of this method requires a knowledge of the in-situ stresses and preconsolidation pressure values; high quality oedometer tests are essential.

3.5.2 SHANSEP

Stress History And Normalized Soil Engineering Properties (SHANSEP) is the basis of the technique. It consists of evaluating the stress history of the clay deposits by evaluating profiles of vertical effective stress (σ'_v) and preconsolidation pressure (σ'_p) to determine OCR profiles through the deposit and then applying the appropriate normalized values to give the representation of strength properties for design.

The basic steps are as follows (Ladd and Foot, 1974; Jamiolkowski et al., 1985):

1. Examine and subdivide the soil profile into component deposits on the basis of boring logs, visual classification, etc.
2. Obtain good "undisturbed" samples and investigate the stress history of the soil profile using a program of total unit weight, pore pressure and vertical effective stress measurements. Check whether or not normalized behavior applies to the soil studied.
3. Perform a series of CK_0U (one-dimensionally consolidated) shear tests on specimens consolidated beyond the in-situ preconsolidation pressure (to σ'_v greater than 2 times σ'_p) to measure the behavior of normally consolidated clay and also on specimens rebounded to different OCR's to measure overconsolidated behavior.
4. Express the results in terms of normalized soil parameters (NSP) and establish NSP vs. OCR relationships, e.g. c_u/σ'_v vs. OCR.

The resulting SHANSEP design strength parameters are expressed in terms of:

$$\frac{c_u}{\sigma'_v} = S\ OCR^m \qquad (3.12)$$

in which S and m vary with the in-situ mode of failure as will be seen in the next sections.

Ladd (1991) made some recommendations for the assessment of S and m. He stated that CL and CH clays (in the unified system of soil classification) tend to have lower, less scattered undrained strength ratios than soils plotting below the A-line. Moreover, he concluded the following range of typical values, based on experience:

- Sensitive marine clay ($PI < 30, IL > 1$): $c_u/\sigma'_p = 0.20$, with a nominal standard deviation (SD) of 0.015 and $m = 1$.
- Homogeneous CL and CH sedimentary clay of low to moderate sensitivity (PI = 20%–80%): $S = 0.20 + 0.05PI$, or simply $S = 0.22$. Moreover, $m = 0.88(1 - C_s/C_c) \pm 0.06SD$, or simply $m = 0.8$.
- Northeastern varved clay: $S = 0.16$ and $m = 0.75$.
- Sedimentary deposits of silts and organic soils (Atterberg limits plot below the A-line, but excluding peats) and clay with shells: $S = 0.25$ with a nominal $SD = 0.05$. Moreover, $m = 0.88(1 - C_s/C_c) \pm 0.06SD$, or simply $m = 0.8$.

Jamiolkowski et al. (1985) stated that there exist evidence to support that for low OCR inorganic clay of low to moderate plasticity, S falls in a very narrow band of 0.23 ± 0.04.

3.5.3 Discussion

The SHANSEP technique has been used in a number of projects (e.g. Koutsoftas, 1981; Koutsoftas and Ladd, 1985; Gaberc, 1994; Lechowicz, 1994).

Koutsoftas and Ladd (1985) studied the suitability of the method and compared it to the conventional practice at that time. They carried out engineering studies for the design of the foundation of an offshore structure on marine clay.

The conventional practice of testing included Unconsolidated Undrained triaxial compression tests to obtain the initial c_u profile and isotropically consolidated-undrained triaxial compression tests to predict strength increase with consolidation. The authors concluded the following:

- Predictions of the initial c_u profile based on Unconsolidated Undrained compression data can easily be in error by ±30%.
- Predictions of the rate of strength gain with consolidation, based on isotropically consolidated triaxial tests, will usually be unsafe by $30 \pm 10\%$.
- When the project does not warrant development of anisotropic strength parameters, design values of c_u can be obtained from one-dimensionally consolidated direct simple shear tests or from equation 3.2 assuming $S = 0.22 \pm 0.03$ and $m = 0.8$. Both being reasonable estimates.
- The SHANSEP technique provides a good or slightly conservative indication of stability, whereas, the conventional practice was seen to be erratic in its prediction of stability and frequently would lead to inadequate designs of either an unsafe or overconservative nature.

Nevertheless, there also are different opinions about the reliability of the method especially when structured soils are tested. Tavenas et al. (1987) state that the structure of the clay is destroyed when the yielding surface is shifted, that is, when the effective stress exceeds the preconsolidation pressure.

The SHANSEP technique, in an attempt to overcome the effect of sampling disturbance, consolidates the samples to stress levels higher (2.5 to 4 times) than those in-situ. Therefore, compared with actual behavior, the SHANSEP approach would thus give a smaller stiffness and shear strength. However, when dealing with "young" soil, where structure is not well developed, the approach should be more appropriate.

Another limitation of the method is that it can only be applied to fairly regular deposits for which a well defined stress-history can be obtained. Ladd and Foot (1974) recommend not to employ the technique when highly heterogeneous deposits are encountered; in that case, the study should be supported by additional field testing.

3.6 Undrained strength anisotropy

The behavior of soil deposits can be anisotropic because of either *structural anisotropy* or *stress induced anisotropy*. Natural, sedimentary, clays are usually structurally anisotropic due to the manner of soil deposition during the formation process; particles tend to become oriented in the horizontal direction during one dimensional deposition. However, macroscopic variations in fabric may also produce inherent anisotropy (e.g. stiff fissured clay, varved clay, etc.).

Early research on this field attempted to evaluate structural anisotropy by testing samples cut at different orientations (β) to the vertical. Figure 3.8 illustrates the anisotropic nature of the undrained strength measured from Unconsolidated Undrained triaxial tests on high quality samples. Since all specimens were sheared in a conventional triaxial apparatus along similar effective stress paths, the reduction in strength compared to vertical loading can be attributed to a preferred particle orientation (structural anisotropy alone).

Soils can also exhibit a stress induced anisotropy whenever K_0 is not equal to unity (K_0-consolidation). A structurally isotropic material may show stress induced anisotropy. It arises from a difference in the normal stress acting in various directions as a result of either special loading conditions or boundary conditions. With reference to triaxial testing, we can state that isotropic consolidation produces stiffening in all directions, while, anisotropic consolidation produces stiffening in a preferred direction. Consequently, the soil will respond differently to loads applied in distinct directions, i.e. compression or extension.

In practice one deals with both anisotropy components simultaneously. The practical significance of anisotropy on staged construction methods is illustrated in figure 3.9 by considering a long embankment constructed on soft clay.

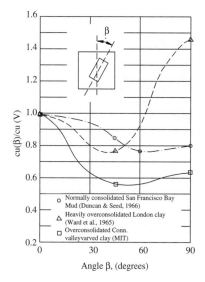

Figure 3.8. Undrained strength of three clays measured by UU triaxial compression (Ladd et al., 1977)

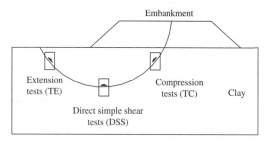

Figure 3.9. Relevance of shear tests to shear strength in the field

Numerical studies also show that foundation clays invariably experience different stress conditions along the potential failure surface (Jardine and Menkiti, 1999) due to principal axes rotation. The stress conditions might be represented by triaxial compression (TC), direct simple shear (DSS) and triaxial extension (TE) at particular points.

The assumption that c_u is unique function of water content has been disproved by the measurement of significant strength anisotropy in clay deposits. For homogeneous non-layered clays sheared under the conditions such as those in figure 3.9, it is typically found that c_u from compression tests (TC) is greater than the value from direct simple shear tests (DSS), which in turn is greater than the value from extension tests (TE).

Figure 3.10 summarizes values of c_u/σ'_v measured in K_0-consolidated specimens in undrained triaxial compression (TC), extension (TE), and in direct simple shear (DSS) for different clays.

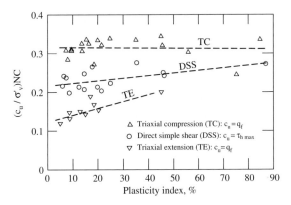

Figure 3.10. Undrained strength anisotropy from CK_0U tests on NC clay (Jamiolkowski et al., 1985)

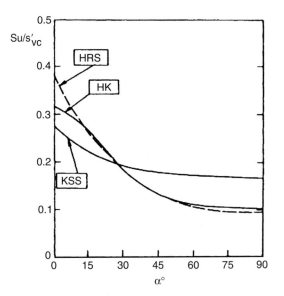

Figure 3.11. Anisotropy of c_u/σ'_v for different soil at OCR=1 (Jardine and Menkiti 1999)

Another means to study anisotropy is through the Hollow Cylinder Apparatus (HCA) which has shown to be an excellent tool of anisotropy investigation as it allows the control of the principal stress orientation.

Jardine and Menkiti (1999) reported results of their experiences with the HCA on three silty soils (denoted by HRS, HK, KSS). Figure 3.11 reveals that the reduction of the undrained strength ratio (S) with the orientation of principal stresses ($\alpha = 0°$ for compression and $\alpha = 90°$ for extension) may be severe in some soils.

From figures 3.10 and 3.11 we can conclude the following:

- Less plastic, and often more sensitive, clays tend to have higher strength anisotropy than more plastic clays.
- The use of an undrained strength ratio estimated from compression tests (CK_0U) in stability calculation will yield unsafe results for clays of low to moderate plasticity index and OCR.

It has been suggested (Jamiolkowski et al., 1985) that most stability analysis should consider strength anisotropy; at least the projects concerning staged construction or unusual loading. Unfortunately, anisotropy cannot be easily characterized and implemented in a model for use in routine work. Still, the assessment of TC and TE tests may help at selecting design parameters even when a simple model (e.g. isotropic) is employed.

3.7 Final recommendations

This section recommends procedures to obtain undrained strength ratio vs. OCR relationships required to calculate the initial c_u profile and subsequent increases due to consolidation. The recommendations are divided into three levels of sophistication, as pointed out by Ladd (1991), depending on the degree of refinement required:

- Level A: For final design of all major projects and for sites where the foundation soils exhibit significant undrained stress-strain-strength anisotropy or contain unusual features (fissuring, varved, highly organic, etc.) and for projects requiring accurate predictions.
- Level B: For preliminary design and for final design of less important projects involving ordinary soils with low to moderate anisotropy.
- Level C: For preliminary feasibility studies and to check the reasonableness of initial strengths inferred from in-situ and laboratory conventional testing programs.

Levels A and B require laboratory CU testing to provide anisotropic and isotropic (average) input strengths, respectively, whereas level C relies on empirical correlations. Table 3.1 summarizes the testing programs recommended by Ladd (1991) from his experiences.

The CK_0U test for Level A use either the SHANSEP technique or simply Reconsolidation to the in-situ stress, depending on the soil type (e.g. highly structured), in-situ OCR and sample quality. The c_u/σ'_v vs. OCR predictions, in the form of equation 3.2, should "exactly" simulate the in-situ response for the first stage of construction, but they may involve errors on the safe side when used to compute strength increases during consolidation.

For Level B programs, Ladd (1991) recommends the use of either CK_0U direct simple shear or CK_0U triaxial compression and extension to estimate

Table 3.1. Recommended laboratory testing program (Ladd, 1991)

Level A	Level B	Level C
CK_0U tests with different modes of failure: – Triaxial compression (TC) – Direct simple shear (DSS) – Triaxial extension (TE)	CK_0U tests with either: – Direct simple shear (DSS) or – Triaxial compression (TC) and Triaxial extension (TC) in order to estimate avrg. strength	Uses empirical correlations rather than testing. See section 4.5 for typical values.

a reasonable average value of shear strength along the potential failure surface of a slope. Moreover, he states that isotropic strength profiles suffice for the assessment of stability. Level B should not rely on tests performed on isotropically consolidated specimens. Level C should only rely on empirical correlations.

Levels A, B and C require a careful assessment of the stress history of the foundation soil. This fact, plus the observation that c_u/σ'_v vs. OCR for most homogeneous soils falls within a fairly narrow range, means that consolidation testing usually represents the single most important experimental component for the design of staged construction projects.

4

Discussion on slope stability evaluation

W.F. Van Impe & R.D. Verástegui Flores
Laboratory of Geotechnics, Ghent University, Belgium

4.1 Preamble

The objective of the present chapter is to review slope stability methods and related issues. The methods have been classified here into two categories: limit equilibrium methods and strength reduction methods.

The assessment of the stability of slopes remains a challenging task of geotechnical engineering. However, many aspects have been thoroughly studied over the last decades and today the methods of analysis are able to tackle complex problems.

Slopes, natural or man-made, are observed to collapse in different ways. Figure 4.1 summarizes some of the most common patterns of soil slope failures. Rocks and soft rocks slopes show different patterns out of the scope of this book.

The two major types of slides are rotational slides and translational slides (Fig. 4.1). Rotational slide are those in which the surface of sliding is curved concavely upward and the slide movement is roughly rotational about an axis parallel to the ground surface. On the other hand, a translational slide is one in which a soil mass moves along a roughly planar surface with little rotation. Such planar movement could be the result of the presence of a weak layer or an interface of different soil types.

Moreover, earthflows and creep are patterns observed especially in soft fine grained soils. Earthflows have a characteristic shape. They occur for example when the slope material liquefies and runs out forming a depression at the head and a mound at the toe. On the other hand, creep manifests as a imperceptibly slow, steady downward of the slope caused by for example, the environment action, presence of existing sliding surfaces and vicinity of stress state to failure.

From these 4 slope failure types, the rotational slide and translational slide were explicitly studied in classic soil mechanics by means of limit states methods. A short review is given in the next sections. Moreover, the stability analysis of quick clay masses is evaluated in detail.

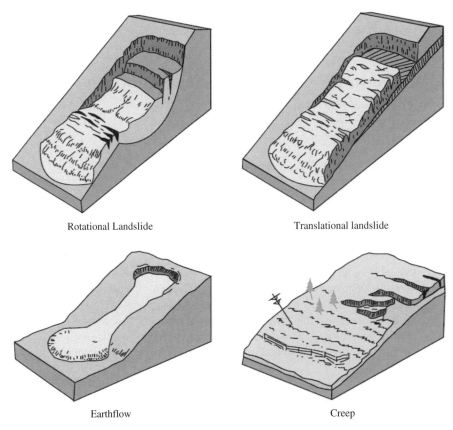

Figure 4.1. Common patterns of soil slope failure (source USGS)

4.2 Causes of slope instability

When facing a design task it is important to understand the causes of instability of a slope to anticipate the changes in the properties of the soil that may occur over time, loading conditions, seepage conditions to which the slope will be subjected, etc.

As stated by Duncan and Wright (2005), when discussing the causes of slope failure it is useful to start from the very fundamental premise that *the shear strength of the soil must be greater than the shear stress required at equilibrium*. Consequently, the most fundamental cause of instability is that for some reason, the shear strength of the soil is less than the shear strength required for equilibrium and such condition can be reached in two ways:

- Through a decrease of shear strength in the soil
- Through an increase of the shear stress required for equilibrium

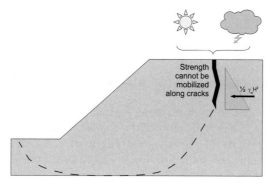

Figure 4.2. Cracking as an effect of environment action and its implication in slope stability

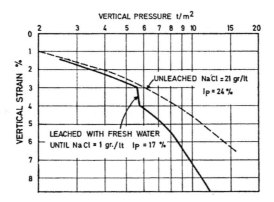

Figure 4.3. Results from fresh water leaching tests on clay specimens from Norway (Bjerrum, 1967)

Reasons for a decrease in the shear strength of the soil are for example an increase of pore water pressures (e.g. due to rainy seasons), cracking (i.e. due to the action of the environment and tension stresses, Fig. 4.2), swelling, leaching (Fig. 4.3), strain softening behavior, cyclic loading (e.g. leading to liquefaction). Figure 4.3 illustrates a striking example of soil behavior. These are results of leaching tests after Bjerrum (1967); they showed at that time how the structure of a marine deposited soil could significantly collapse when it is leached with fresh water creating sensitive clays. This particular example should always remind the engineer to stay alert and pay attention to the various ways of soil behavior. Special considerations on the analysis of quick clays are given in section §4.5.

Reasons for an increase of the shear stress required for equilibrium are for example an extra loading acting on the slope, water accumulation in cracks (Fig. 4.2), increase of the unit weight of the soil (e.g. due to wetting), excavation works at the toe of the slope, drop in water level at the site (e.g. due to water pumping), earthquake or other type of dynamic loading, etc.

In reality, slopes will fail usually because of a combination of some of the reasons cited above.

4.3 Stability conditions for analysis

The first requirement to perform slope stability analysis is to formulate correctly the problem. Selecting appropriate conditions for analysis of slopes requires considerations of the shear strength of soils under drained and undrained conditions, or under drainage conditions that will occur in the field.

The general principles involved in selecting analysis conditions and shear strengths are summarized in table 4.1.

When an embankment is constructed on a clay foundation, the embankment load causes the pore water pressure in the clay to increase. After a period of time, such increment will gradually dissipate and eventually the pore water pressures will return to the initial steady value. As the excess pore water pressure dissipates, the effective stresses in the foundation soil increase, the strength of the clay increases and as a result the factor of safety increases too. Figure 4.4 illustrates these relationships and out of it one may conclude that the most critical condition occurs at the end of construction (undrained). Then, it is only necessary to analyze the end-of-construction condition.

Table 4.1. Shear strength for stability analysis (Duncan, 1996)

	Condition		
	End of construction	Staged construction	Long term
Procedure and strength for sand	Effective stress anal. with c' and ϕ'	Effective stress anal. with c' and ϕ'	Effective stress anal. with c' and ϕ'
Procedure and strength for clay	Total stress anal.	Total stress anal. with c_u from approp. consolidation anal.	Effective stress anal. with c' and ϕ'

When a slope in clay is created by excavation, the pore pressures in the clay decrease in response to the removal of the excavated material. Over time, the negative excess pore water pressure dissipate and the pore pressures eventually return to the initial steady value. As the pore water pressure increases, the effective stress in the decreases and the factor of safety decreases with time as illustrated in figure 4.4. Out of these relationships it can be concluded that the long-term (drained) condition is more critical than the end-of-construction condition.

Drained conditions are analyzed in terms of effective stresses using values of c' and ϕ' determined from drained tests, or from undrained tests with

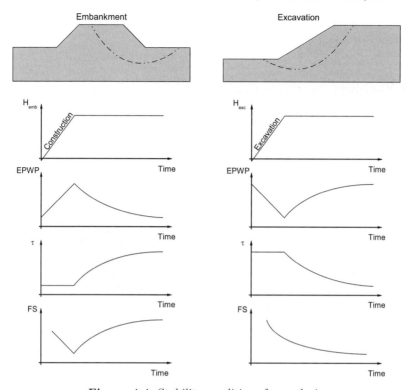

Figure 4.4. Stability conditions for analysis

pore water pressure measurements. When dealing with clay, drained triaxial tests are frequently impractical because the required time is very long, therefore, undrained triaxial tests with pore pressure measurements are the most suitable. Values of ϕ' for natural deposits of cohesionless soils are usually estimated using correlations from field tests results (i.e. CPT, SPT, DMT, PMT) given the current difficulties of testing high quality undisturbed sand samples.

Undrained conditions are analyzed in terms of total stress in order to avoid having to rely on estimated values of pore water pressure for undrained loading conditions. Undrained shear strength of soils is usually correlated from field tests or laboratory tests. For staged construction analysis, the undrained strength is furthermore assessed through consolidation analysis in combination with for example triaxial CU testing.

In cases where it is not clear whether the short-term or long-term condition will be more critical, both should be analyzed, to ensure that the slope will have an adequate stability under any condition.

Another important topic is that of the selection of an suitable factor of safety for design. The main considerations to take into account are the degree of uncertainty in evaluating conditions and shear strengths for analysis and the possible consequences of failure. Typical minimum acceptable values of factor of safety are about 1.3 for the categories end-of-construction and staged-construction and 1.5 for long-term conditions.

4.4 Stability analysis procedures

The universal availability of computers and a much improved understanding of the mechanics of slope stability analysis have brought about considerable changes in the computational aspects of slope stability analysis in the last years.

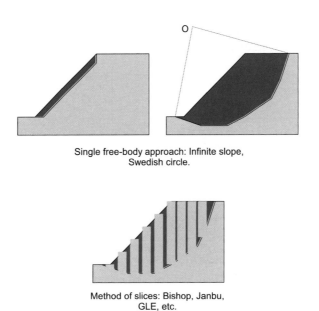

Single free-body approach: Infinite slope, Swedish circle.

Method of slices: Bishop, Janbu, GLE, etc.

Figure 4.5. Limit equilibrium approaches

4.4.1 Limit equilibrium methods

The calculation of global stability is commonly expressed in terms of the Factor of Safety computed by means of *Limit Equilibrium Methods*. The principles underlying these methods are as follows:

- A slip mechanism is postulated (i.e. a potential failure surface is outlined).
- The shearing stress to equilibrate the assumed slip mechanism is calculated by means of statics.
- The calculated shearing stress required for equilibrium is compared to the available shear strength in terms of the factor of safety.
- The mechanism with the lowest factor of safety is found by iteration.

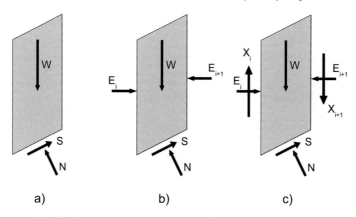

Figure 4.6. Free-body equilibrium of slices: a) ordinary method of slices, b) simplified Bishop procedure, c) more refined methods

In general, two approaches to study the statics of limit equilibrium methods can be identified: the single free-body approach and the method of slices. The single free-body approach considers one single free body that slides downward. The infinite slope and the Swedish circle are examples of such analysis (Fig. 4.5) although their applicability is limited in practice. On the other hand, the method of slices includes such methods as the Ordinary Method of Slices, Simplified Bishop procedure and others. Unlike single-free body analysis, it subdivides the soil mass prone to sliding in slices and considers the equilibrium of each of them. Figure 4.6 illustrates the forces acting on the slices according to some methods. The simplest methods do not consider any interslice forces while other more elaborated do.

Some of the method of slices proposed in literature and their satisfied equilibrium conditions are summarized in table 4.2.

4.4.2 Strength reduction methods

More recently, the following definition of factor of safety has gained a lot of acceptance: *The Factor of Safety is that factor by which the shear strength parameters may be reduced in order to bring the slope into a state of failure.* The preceding definition has given rise to a new technique, the so called *Strength Reduction Methods* (SRM). Such definition has been easily implemented on finite element and finite difference computer programs (i.e. PLAXIS, FLAC).

The principle behind the shear strength reduction technique in finite element slope stability analysis is to reduce c' and ϕ' or c_u by a factor until failure occurs. Consequently, the overall safety factor and the corresponding potential failure surface can be obtained simultaneously. Finite element slope

Table 4.2. Characteristics of equilibrium methods of slope stability analysis (Duncan, 1992)

Method	Characteristics
Slope stability charts	Fast and accurate enough for many purposes
Ordinary method slices	For circular slip surfaces
	Satisfies moment equilibrium only
Bishop's modified method	For circular slip surfaces
	Satisfies moment equilibrium
	Satisfies vertical forces equilibrium only
Force equilibrium method	For any shape of slip surface
	Does not satisfy moment equilibrium
	Satisfies vert. and horiz. force equilibrium
Morgenstern and Price	For any shape of slip surface
	Satisfies all equilibrium conditions
Spencer's method	For any shape of slip surface
	Satisfies all equilibrium conditions

failure prediction by the shear strength technique is performed by using two reduced shear strength parameters, namely:

$$c_R = \frac{c}{R} \quad (4.1)$$

$$\tan \phi_R = \frac{\tan \phi}{R} \quad (4.2)$$

where R is called the shear strength reduction factor.

A starting value of 1 (no reduction of strength) is usually given to R. As the calculation proceeds, R is increased and the shear strain and displacements are evaluated for each step until failure is reached. The shear strength factor (R) at failure is called critical strength reduction factor (R_f) and corresponds to the overall safety factor of the slope.

The output of the calculation is expressed as a graph illustrating the progress of R. Figure 4.7 shows the development of the strength reduction factor (R) versus the displacement of a control node within a finite element grid modeling a slope stability problem (Fig. 4.8).

Note that it is only meaningful to refer to a factor of safety when a steady state solution is obtained. Intermediate values of R do not have any physical meaning and are only used for numerical purposes. The displacements obtained during the calculations also do not have any physical relevance. So, by looking at the progress of calculation for a particular slope (Figs 4.7 and 4.8) it can be deduced that a steady state solution has been clearly obtained at the end of the calculation as indicated by the flat slope of the curve. The factor of safety is therefore (for this example) determined as 1.6 approximately.

Although the magnitude of total displacements does not have physical meaning, the displacement pattern calculated for the last step (incremental

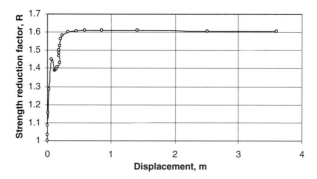

Figure 4.7. Strength reduction factor (R) versus displacement.

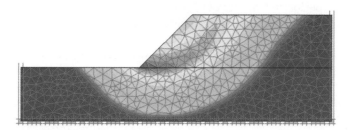

Figure 4.8. Shadings of incremental displacements delineating the failure mechanism (PLAXIS output)

displacements) provides an idea of the failure mechanism developing in the slope. Note that the failure mechanism, as shown in figure 4.8 is not fully circular.

The shear strength reduction technique has a number of advantages over the limit equilibrium method (Matsui and Ka-Ching, 1992). Probably the most remarkable advantage is the automatic failure surface determination. The application of this technique has been limited in the past due to the long computer run times required. But with the increasing speed of computations, the use of the technique is also increasing.

In order to investigate the results of the strength reduction method, factors of safety obtained with PLAXIS have been compared to factor of safety estimated by the limit equilibrium method. The next section describes this comparative analysis in more detail.

4.4.3 Limit equilibrium versus strength reduction methods

A number of simulations were performed for a wide range of parameters and for different embankment geometries with a variety of slopes ranging from 15° to 90°. The results were back calculated to obtain coefficients or stability factors from methods generally accepted such as the Taylor method (Taylor,

4.4.3.1 Taylor's method

Taylor (1937) defined the stability factor N_0, as shown by equation 4.3, to prepare a chart in order to determine the stability of slopes in a homogeneous deposit of soil underlain by a much stiffer strata.

$$N_0 = FS\frac{\gamma H}{c_u} \qquad (4.3)$$

Equation 4.3 was used to estimate the Taylor's stability factor (N_0) from simulations performed in PLAXIS given the unit weight (γ), the embankment slope (H), the undrained strength (c_u) and the factor of safety (FS) obtained.

The back calculated stability factor (N_0) was then plotted against the slope inclination. Figure 4.9 shows the stability chart as proposed by Taylor (1937) and the back calculations from PLAXIS. One can see that the approach is close and the trends are similar.

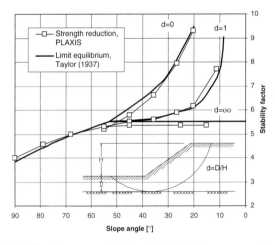

Figure 4.9. Total stress analysis; stability factor versus slope angle

The points calculated through the strength reduction method are within a few percent of the limit equilibrium solution. Nonetheless, one must bear in mind that Taylor's solution refers exclusively to circular surfaces while the strength reduction method does not impose any restriction to the geometry of the failure mechanism. Therefore, the slight differences might be reflecting the limitations of Taylor's method instead of the incorrectness of the strength reduction method.

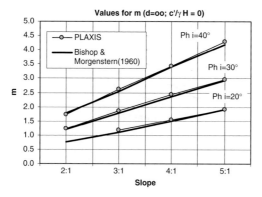

Figure 4.10. Effective stress analysis; stability coefficient m for $c/\gamma H = 0$ and $d = \infty$

4.4.3.2 Bishop-Morgenstern method

Bishop and Morgenstern (1960) prepared a number of charts for homogeneous soil slopes with simple geometry using Bishop's simplified method of slices. They expressed the factor of safety as follows:

$$FS = m - nr_u \qquad (4.4)$$

where m and n are stability coefficients that depend on the drained friction angle, the drained cohesion (if exists) and the geometry of the slope. The procedure also assumes circular potential failure surfaces.

In view of the amount of charts for the method, the analysis carried out here has been restricted. Only 2 charts were employed here to compare the results; one of them with $c' = 0$ and the other with $c' = 0.05H$. Moreover, absence of pore water pressure was adopted ($r_u = 0$). In that way, the factor of safety calculated with PLAXIS directly equals to m.

Figures 4.10 and 4.11 show the charts for estimation of the stability coefficient m proposed by Bishop and Morgenstern (1960) and the back calculated points from PLAXIS output. One can see that the agreement is much closer.

Finally, it can be concluded that Strength reduction factors of safety were within a few percent of the limit equilibrium solution and that in general a close agreement was observed. The fact that the limit equilibrium methods employed here (e.g. Taylor, 1937; Bishop-Morgenstern, 1960) assume circular surfaces of potential failure may be causing such small deviations.

4.5 Failure mechanisms for highly sensitive clays – Van Impe and De Beer (1984)

Highly sensitive clays or quick clays do relate to clays that when remolded loose their structure completely and then their shear strength (after remolding) is reduced almost to zero.

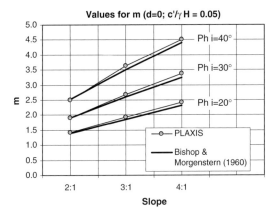

Figure 4.11. Effective stress analysis; stability coefficient m for $c'/\gamma H = 0.05$ and $d = 1$

Sensitivity is the ratio between the undisturbed and the fully remolded undrained shear strength. In this framework, quick clay can be defined as a clay with a sensitivity of 50 or more and a fully remolded shear strength of less than 0.4 kPa.

Quick clays are found in areas once glaciated during the Pleistocene epoch. They have mainly been located in northern Russia, Norway, Finland, Sweden, Canada and Alaska. These areas are all characterized by geological isostatic uplift which took place after the retreat of the ice. The development of very high sensitivity is often the result of processes that have taken place after the deposition of the clay layers such as leaching of salts (remotion of dissolved salts), changes in the ion composition of the pore water, pH of the pore water, dispersive action of some organic and inorganic natural compounds, etc.

The slope stability analysis in highly sensitive clays as proposed by Norwegian researchers (Bjerrum, 1973; Gregersen, 1981; Aas, 1981) is mainly based on the drained shear behavior of the soil in or along the the potential sliding surfaces. It has been reported in involved Norwegian investigations that immediately after the sliding of a quick clay mass, no excess pore water pressures are measured in the immediate vicinity of the sliding surface. This is often considered to be an experimental support for using drained parameters in the stability analysis.

However, the possibility of the existence of excess pore water pressures in the more pervious silt seams of the undisturbed quick clay layer cannot be excluded. Such excess pore water pressures can disappear as soon as the sliding occurs. They however can be generated simultaneously in the quick clay itself, due to remolding during sliding.

In the Swedish approach, as described by Bernander (1981), use is made of the undrained shear strength parameters. It is assumed that the shear plane coincides with the interface of the metastable quick clay layer and the more resistant substratum. The mentioned contribution stresses the large influence of the configuration of that contact plane on the stability conditions. In situ observations show the sliding does not always occur along the contact plane with the substratum, but also other shear planes are frequently observed.

Analyzing slidings of quick clays, initiated apparently without any external cause, Bernander concludes some of these slidings must be related to creep phenomenae described as proposed by Mitchell et al. (1968). If such creep phenomenae however only should be a consequence of the shape of the contact plane itself, they should have started from the very beginning of the formation of the actual geometry of the soil surface and of the involved layers. Such creep phenomenae consequently exist for several decades and deformation velocity therefore should be extremely small.

Tests results of Bernander (Fig. 4.12) indicate, for a clay at a given overconsolidation ratio, that the ratio of the residual shear strength τ_r at large deformation (δ) to the maximum shear stress (τ_{max}) tends toward unity when the deformation speed (v) decreases significantly. Therefore, it seems unlikely the creep phenomenae mentioned before should be able to explain the occurrence of a landslide after several centuries of existence of the clay layer interface in its actual state.

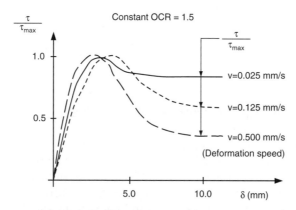

Figure 4.12. Stress strain curves of clay specimens sheared at different strain rates (after Bernander, 1981)

On the contrary, the activation of creep due to some new external factors can contribute to such explanation.

Moreover, from literature in general it seems that insufficient attention is paid to the possible existence of excess pore water pressures developed in the silty seams often detected in such clay masses.

4.5.1 Flake type sliding of quick clay

In order to explain flake type sliding of quick clay masses, the reasoning as proposed by Aas (1981) is commented first. For sake of simplicity the case of

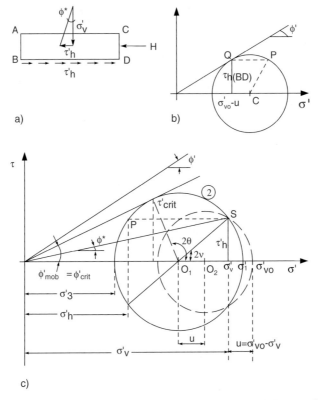

Figure 4.13. Stress state at a horizontal potential sliding surface

horizontal soil surface and horizontal potential sliding plane BD is considered (see Fig. 4.13a). In the original situation at rest, the effective stresses $\sigma'_{v,0}$ and $K_0 \cdot \sigma'_{v,0}$ are assumed in the vicinity of the plane BD. In addition, by applying an horizontal external force H, shear stresses τ_h are introduced.

From the Mohr circle 2, in figure 4.13b, representing the stress state in the plane BD, one gets:

$$\tau_h = \sqrt{\left(\frac{\sigma'_1 - \sigma'_3}{2}\right)^2 - \left(\sigma'_{v,0} - u - \frac{\sigma'_1 + \sigma'_3}{2}\right)^2} \quad (4.5)$$

and

$$\frac{\sigma'_1 + \sigma'_3}{2} = \frac{\sigma'_v + \sigma'_h}{2} = \frac{\sigma'_{v,0} - u + K_0 \sigma'_{v,0} - u}{2} \quad (4.6)$$

or

$$\frac{\sigma'_1 + \sigma'_3}{2} = \sigma'_{v,0}\frac{1 + K_0}{2} - u \quad (4.7)$$

$$\sigma'_{v,0} - u - \frac{\sigma'_1 + \sigma'_3}{2} = \sigma'_{v,0}\frac{1 - K_0}{2} \quad (4.8)$$

If the angle of tangent through the origin at the Mohr circle 2 is designated ϕ'_{mob}, the following can be written:

$$\frac{\sigma'_1 - \sigma'_3}{2} = \frac{\sigma'_1 + \sigma'_3}{2} \sin \phi'_{mob} \tag{4.9}$$

or

$$\frac{\sigma'_1 - \sigma'_3}{2} = \left(\sigma'_{v,0} \frac{1+K_0}{2} - u\right) \sin \phi'_{mob} \tag{4.10}$$

and therefore:

$$\tau_h = \sigma'_{v,0} \cdot \sqrt{\left(\frac{1+K_0}{2} - \frac{u}{\sigma'_{v,0}}\right)^2 \sin^2 \phi'_{mob} - \left(\frac{1+K_0}{2}\right)^2} \tag{4.11}$$

The equation 4.11 has a general validity. From it, the variation of excess pore water pressure with $\sin \phi'_{mob}$, when $\sigma'_{v,0}$ and τ_h are given values, can be deduced:

$$\frac{\Delta}{\Delta \sin \phi'_{mob}} \left\{ \left(\frac{1+K_0}{2} - \frac{u}{\sigma'_{v,0}}\right)^2 \cdot \sin^2 \phi'_{mob} \left(\frac{1+K_0}{2}\right)^2 \right\} = 0 \tag{4.12}$$

Starting with the value of $u = 0$ at the initial stress conditions, one obtains from equation 4.12:

$$\frac{\Delta \frac{u}{\sigma'_{v,0}}}{\Delta \sin \phi'_{mob}} = \frac{1+K_0}{2 \sin \phi'_{mob}} \tag{4.13}$$

On figure 4.14b after Aas, from CU triaxial tests performed on five different normally consolidated clays, the results are shown of the variation of a pore water pressure function $F(u)$ versus the values of the mobilized angle ϕ'_{mob} of internal friction.

The curves with respect to the indicated numbers 4 and 5 for non-sensitive clays are from a certain level for $\sin \phi'_{mob}$ quite different from those with respect to the sensitive clays (numbers 1, 2 and 3). For the last mentioned, starting at values $\sin \phi'_{mob} \approx 0.42$ to 0.45, a marked rising of the pore water pressure, and a corresponding decrease of the shear strength (Fig. 4.14a) is found.

After reaching the critical ϕ'_{mob} value of the angle of friction, the shear strength of the quick clay drops drastically (see Fig. 4.14a) because the further rising up of the mobilized friction angle ϕ' cannot compensate the decrease of the effective stress σ' caused by the rising of the excess pore water pressure u. On the contrary, that is not the case for two non-sensitive clays (4 and 5).

So, if for quick clays along whatever plane, the critical value of ϕ' is mobilized, any small increase of the shear stress will cause an almost total reduction of the shear strength.

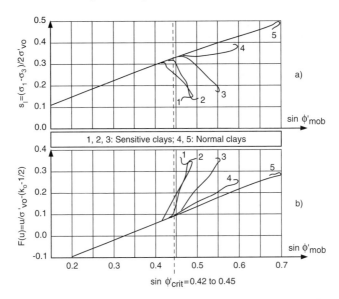

Figure 4.14. Variation of shear strength and pore water pressure function $F(u)$ with mobilized shear angle

When for kinetic reasons only a continuous sliding along the horizontal plane BD should be possible, the point representing the effective stress on that horizontal plane in the Mohr circle (for normal clays at sliding conditions) should be located on the intrinsic straight line OQ at a friction angle ϕ' (Fig. 4.13c). When the values of ϕ' and τ_h are given, the possibility of sliding along the horizontal plane needs the development of an excess pore water pressure u given by:

$$\tau_{hBD} = \left(\sigma'_{v,0} - u\right) \cdot \tan \phi' \qquad (4.14)$$

In quick clays however the conditions for continuous sliding along the horizontal plane BD are much less severe. Indeed, when in such clays along an arbitrary plane (not necessarily a kinematically possible), the critical value $\phi'_{crit} = \phi'_{mob} < \phi'$ of the friction angle is mobilized, any very small increase of the shear stress will cause a drastic decrease of the shear strength and a subsequent liquefaction.

From the previous considerations it results for quick clays the Mohr circle 2 in figure 4.13b representing a critical stress state. The ratio of the shear stress τ_h to the vertical effective stress σ'_v on the sliding plane (Fig. 4.13b) is defined by an apparent angle ϕ^*, much smaller than ϕ'_{mob}.

4.5.2 Analysis taking pore water pressures into account

Suppose a mass of quick clay with inclined surface at a small angle β and consider a potential sliding plane parallel to the mentioned ground level

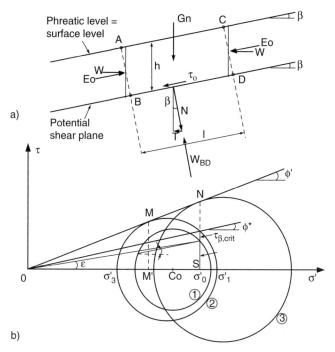

Figure 4.15. Stress state of a inclined slope with water level at ground surface

(Fig. 4.15a). The phreatic level in each point is thought to correspond with the ground surface, then a groundwater flow parallel to the surface takes place. When the slope is infinitely extended, each vertical section is a symmetrical one and the indicated forced W and E_0 are independent of the considered section.

For the quick clay soil mass ABCD of figure 4.15a, one gets a tangential force T:

$$T = G_n \cdot \sin \beta \tag{4.15}$$

or

$$T = \gamma_s \cdot h \cdot l \cdot \sin \beta \cos \beta \tag{4.16}$$

where γ_s is the unit weight of the soil.

So a shear stress value τ_0 is working on BD:

$$\tau_0 = \gamma_s \cdot h \cdot \cos \beta \sin \beta \tag{4.17}$$

The total normal force N on BD is given by:

$$N = \gamma_s \cdot h \cdot l \cdot \cos^2 \beta \tag{4.18}$$

Taking into account the uplift force of the water:

$$W_{BD} = \gamma_w \cdot h \cdot l \cdot \cos^2 \beta \tag{4.19}$$

The initial effective stress σ_0' on the plane BD becomes:

$$\sigma_0' = (\gamma_s - \gamma_w) \cdot h \cdot \cos^2 \beta \tag{4.20}$$

For the normally consolidated sensitive clay, the equilibrium conditions may be written as:

$$\tau \leq \sigma_0' \cdot \tan \phi' \tag{4.21}$$

Rewriting equation 4.21, considering the expressions 4.17 and 4.20, this equilibrium condition is given by:

$$\tan \beta \leq \frac{\gamma_s - \gamma_w}{\gamma_s} \tan \phi' \tag{4.22}$$

It can be seen from figure 4.15b, the resulting effective stress on BD is inclined over an angle ϵ toward the normal on BD;

$$\tan \epsilon = \frac{\tau_0}{\sigma_0'} = \frac{\gamma_s}{\gamma_s - \gamma_w} \tan \beta \tag{4.23}$$

About the value of the angle of internal friction ϕ' of the undisturbed soil skeleton to be considered, extensive research by means of different kinds of laboratory tests was performed (Bjerrum, 1969; Aas, 1981). Out of the Norwegian test results it seems the ϕ' value to be taken into account for the shear strength of undisturbed Norwegian quick clay skeleton is about $\phi'_{mob} = \phi' \approx 25°$, that is $\sin \phi'_{mob} \approx 0.42$ (see Fig. 4.14).

If, for example, $\gamma_s = 18.5\,\text{kN/m}^3$ and $\gamma_s - \gamma_w = 8.5\,\text{kN/m}^3$, in order to fulfill the equilibrium condition, the angle β (Fig. 4.15a) of the inclination of the sliding surface must be limited to (Eq. 4.22):

$$\tan \beta \leq \frac{8.5}{18.5} \tan \phi' \tag{4.24}$$

or

$$\tan \beta \leq 0.21 \tag{4.25}$$

or

$$\beta \leq 12° \tag{4.26}$$

This means, in the case without considering any pore water pressure development, the slope stability could be guaranteed for a value β up to (Fig. 4.15b):

$$\tan \beta \leq \frac{SN}{SO} = \tan \phi' = 0.466 \tag{4.27}$$

or

$$\beta \leq \phi' \approx 25° \tag{4.28}$$

Thus, if in case of seepage, a surface with a slope of $\beta = 25°$ should be possible; however in case of the existence of a seepage parallel to the ground surface, the slope angle is reduced to $\beta = 12°$. When, however, the water pressures are artesian, the stability conditions can become much more unfavorable.

Let us now consider a quick clay mass with surface inclined at an angle β satisfying the inequality 4.25 and therefore it is in an equilibrium state. On the clay mass supplementary disturbing forces are applied such that, on the most unfavorable elementary plane in the considered point, there exists an effective stress represented by point M on Mohr circle 2 in figure 4.15b is located on the intrinsic line with an angle ϕ'. In normal clays these most unfavorable elementary planes do not constitute a kinematic possible sliding plane, as it is supposed that such a sliding only is possible along a plane BD parallel to the soil surface (Fig. 4.15a). In normal clays the representative point of the maximum shear stress on an elementary plane parallel to BD should be located in point N of Mohr circle 3 in figure 4.15b. In quick clays a distinction has to be made between slow loading and rapid loading. Mohr circle 3 is representative of a slow loading condition, on the other hand, Mohr circle 2 is representative of a rapid loading reaching the critical stress conditions in the elementary plane of the point M where liquefaction is induced. It is consequently sufficient (along the sliding plane BD of Fig. 4.15a) to attain the shear stress SQ in order to have general sliding. The apparent angle ϕ^* can be evaluated as follows:

$$\sigma'_0 \tan \phi^* = \sqrt{\left(\frac{\sigma'_1 - \sigma'_3}{2}\right)^2 - \left(\sigma'_0 - \frac{\sigma'_1 + \sigma'_3}{2}\right)^2} \qquad (4.29)$$

and with

$$\frac{\sigma'_1 - \sigma'_3}{2} = \frac{\sigma'_1 + \sigma'_3}{2} \sin \phi' \qquad (4.30)$$

one gets:

$$\frac{\sigma'_1 + \sigma'_3}{2} = \frac{\sigma'_0}{1 - \sin^2 \phi'} \left[1 \pm \sqrt{1 - (1 - \sin^2 \phi')(1 + \tan^2 \phi^*)}\right] \qquad (4.31)$$

Also:

$$\sigma'_1 = \frac{\sigma'_0 (1 + \sin \phi')}{1 - \sin^2 \phi'} \left[1 \pm \sqrt{1 - (1 - \sin^2 \phi')(1 + \tan^2 \phi^*)}\right] \qquad (4.32)$$

and

$$\sigma'_3 = \frac{\sigma'_0 (1 - \sin \phi')}{1 - \sin^2 \phi'} \left[1 \pm \sqrt{1 - (1 - \sin^2 \phi')(1 + \tan^2 \phi^*)}\right] \qquad (4.33)$$

For the value of $\tau_{crit} = M'M$ (Fig. 4.15b) the following expression can be used:

$$\tau_{crit} = \frac{\sigma'_1 - \sigma'_3}{2} \cos \phi' \qquad (4.34)$$

From several experimental data, mainly based on laboratory tests on Norwegian clays, the NGI deduced:

$$\tan \phi^* = 0.2$$

or

$$\phi^* = 11°18'$$

Introducing in equation 4.31 the values of $\phi' \approx 25°$ and $\phi^* \approx 11°$ one can obtain:

$$\sigma'_1 + \sigma'_3 = \frac{2\sigma'_0}{1 - 0.18}\left[1 - \sqrt{1 - (1 - 0.18)(1 + 0.04)}\right] \tag{4.35}$$

or

$$\sigma'_1 + \sigma'_3 = 1.5\sigma'_0 \tag{4.36}$$

From equations 4.32 and 4.33 we get:

$$\sigma'_1 = \frac{\sigma'_0(1 + 0.42)}{1 - 0.18}\left[1 - \sqrt{1 - (1 - 0.18)(1 + 0.04)}\right] \tag{4.37}$$

or

$$\sigma'_1 = 1.07\sigma'_0 \tag{4.38}$$

For σ'_3 in the same way it is derived:

$$\sigma'_3 = 0.44\sigma'_0 \tag{4.39}$$

The value of $\tau_{crit}^{MM'} = MM'$ is obtained from equation 4.34:

$$\tau_{crit}^{MM'} = 0.285\sigma'_0 \tag{4.40}$$

To this critical shear stress $\tau_{crit}^{MM'}$ on the shear plane corresponds, on the plane BD, a shear stress $\tau_{\beta,crit} = SQ$ (Fig. 4.15b)

$$\tau_{\beta,crit} = SQ = \tan\phi^*\sigma'_0 \approx 0.2\sigma'_0 \tag{4.41}$$

or

$$\tau_{\beta,crit} = 0.2\left(\gamma_s - \gamma_w\right)h\cos^2\beta \tag{4.42}$$

4.5.3 Mechanism of sliding in quick clay masses

4.5.3.1 Disturbing action by disappearance of downward supporting forces

In figure 4.16a a quick clay mass ABCD initially in equilibrium is considered. At the downward end at section AB, suddenly all supporting forces are assumed to disappear. In a quick clay, a sliding will occur when the shear stresses along BD reach the critical value $\tau_{\beta,crit}$, given by:

$$\tau_{\beta,crit} \cdot L = E_a + W + \gamma_s \cdot h \cdot L \cdot \cos\beta\sin\beta \tag{4.43}$$

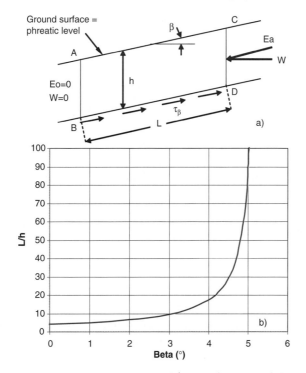

Figure 4.16. Variation of L/h as a function of β

where E_a is the active earth pressure at sliding on plane DC:

$$E_a = K_a \left(\gamma_s - \gamma_w\right) \frac{h^2}{2} = \tan^2\left(\frac{\pi}{4} - \frac{\phi'}{2}\right)(\gamma_s - \gamma_w)\frac{h^2}{2} \qquad (4.44)$$

From equations 4.42 and 4.43 one gets:

$$0.2(\gamma_s - \gamma_w) \cdot h \cdot L \cdot \cos^2\beta = K_a\left(\gamma_s - \gamma_w\right)\frac{h^2}{2} + \gamma_w\frac{h^2}{2} + \gamma_s \cdot h \cdot L \cdot \cos\beta \sin\beta \qquad (4.45)$$

or

$$L = \frac{h(\gamma_s - \gamma_w)K_a + \gamma_w}{2\cos\beta\left[0.2(\gamma_s - \gamma_w)\cos\beta - \gamma_s\sin\beta\right]} \qquad (4.46)$$

For the case of an extremely slow change of the stress conditions in the fully saturated quick clay mass (as shown before in equations 4.22, 4.24 and 4.25) the quick clay could remain in metastable equilibrium for values of the slope angle not higher than $\beta_{\max} = 12°$, for the chosen numerical example. At this extreme value of β any external disturbing force will cause liquefaction of the sensitive clay mass.

On the other hand, in the case of a relatively quick change of the loading, as for example the sudden disappearance of the downward supporting forces,

Discussion on slope stability evaluation

for quick clays the maximum slope angle β_{crit} is limited to (from equation 4.22):

$$\tan \beta_{crit} = \frac{\gamma_s - \gamma_w}{\gamma_s} \tan \phi^* \qquad (4.47)$$

Substituting $\phi^* = 11°18'$, $\gamma_s = 18.5\,\text{kN/m}^3$ and $\gamma_w = 10\,\text{KN/m}^3$ we get:

$$\beta_{crit} = 5°16' \qquad (4.48)$$

As it is probable that quick clay masses were subjected at some point in their geological history to rapid loading conditions, one can expect the natural inclination of the ground level to be not higher than such limiting value of $\beta_{crit} \approx 5°$.

Concerning the variation of the length of the sliding mass as a function of the inclination β, one gets out of equation 4.46 and for assumed values of ϕ^*, γ_s and γ_w:

$$\frac{L}{h} = \frac{8.5 \tan^2\left(45° - 11°18'\right) + 10}{2 \cos \beta \left(0.2 \cdot 8.5 \cdot \cos \beta - 18.5 \sin \beta\right)} \qquad (4.49)$$

The variation of L/h versus β is illustrated by the curve in figure 4.16b. For β values up to about $4°$ the value of L/h is rather small. All of this is based on the assumption of a phreatic level coinciding with the soil surface. For smaller values of β the relative length L/h remains small. In such cases it is probable a succession of retrogressive shell shaped slidings will occur. When on the contrary the slope angle β reaches its critical value β_{crit}, the relative length L/h becomes more significant and a flake type sliding will be more likely to result.

4.5.3.2 Considering excess pore water pressures built up from the upper end in more pervious seams

In many descriptions of quick clay landslides (Aas, 1981; Broms, 1983; Gregersen, 1981) it is mentioned that more pervious seams are often found in the clay mass. For several reasons it can occur that excess pore water pressures at a given moment are developed from the upper region of the slope in the more pervious seams.

For example, in the very well know case of the Rissa landslide, a flake type of sliding occurred after successive shell shaped retrogressive slidings. On the surroundings of the retrogressive slidings high back scarps of several meters were reported.

At the start of a shell shaped slide, it can be assumed that the total weight of the collapsing soil is supported by water, so creating an excess pore water pressure u:

$$u = \gamma_s \cdot h_{shell} - \gamma_w \cdot shell \qquad (4.50)$$

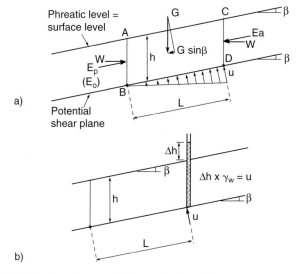

Figure 4.17. Quick clay slope stability analysis taking pore pressure built up into account

This local excess pore water pressure can penetrate into the still remaining clay mass and alter the equilibrium state. In order to get an idea of the possible influence of the excess pore water pressure a very simplified approach can be made.

Let us suppose, at the upper end of the soil mas ABCD (Fig. 4.17), an excess pore water pressure u is built up in the more pervious seams. Assuming a simple triangular distribution of u and from equations 4.42, 4.43 and 4.45 one gets:

$$\tau_{\beta,crit} \cdot L + E_p = E_a + \gamma_s \cdot h \cdot L \cdot \cos\beta \sin\beta \qquad (4.51)$$

where E_p is the passive earth pressure at the lower end of the sliding mass.

When the value of u remains sufficiently below the value of σ'_0, it is possible to write $\tau_{\beta,crit}$ as (from Eq. 4.42):

$$\tau_{\beta,crit} = 0.2\left[(\gamma_s - \gamma_w)h\cos^2\beta - \frac{u}{2}\right] \qquad (4.52)$$

As in the case of quick clays, liquefaction should already have taken place before a passive earth pressure state can be reached; in equation 4.51 E_p is changed in the neutral earth pressure value E_0. From equations 4.51 and 4.52 it results:

$$E_a + \gamma_s hL\cos\beta\sin\beta \geq E_0 + 0.2\left[(\gamma_s - \gamma_w)h\cos^2\beta - \frac{u}{2}\right]L \qquad (4.53)$$

As the angle β is at most $\beta \approx 5°30'$, $\cos\beta$ is put equal to 1. Consequently:

$$\gamma_s hL\sin\beta \geq E_0 - E_a + 0.2\left[(\gamma_s - \gamma_w)h - \frac{u}{2}\right]L \qquad (4.54)$$

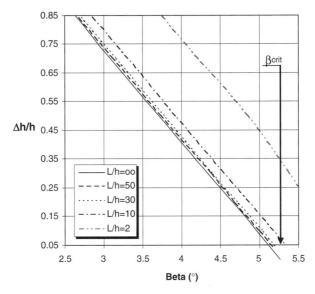

Figure 4.18. Variation of $\Delta h/h$ as a function of β

with

$$E_0 - E_a = \left[(1 - \sin\phi') - \frac{1 - \sin\phi'}{1 + \sin\phi'}\right](\gamma_s - \gamma_w)\frac{h^2}{2} \qquad (4.55)$$

In order to prevent the phenomenon of piping in the clay, the following condition can be written:

$$u < (\gamma_s - \gamma_w)h = 0.85h \qquad (4.56)$$

When $u = \gamma_w \Delta h$, equation 4.56 leads to:

$$\frac{\Delta h}{h} < \frac{\gamma_s - \gamma_w}{\gamma_w} = 0.85 \qquad (4.57)$$

At $\Delta h/h = 0.85$ piping in the clay mass occurs instead of sliding. For all values of $\Delta h/h < 0.85$, the expression of $\sin\beta$ (with assumed values of ϕ', γ_s and γ_w) from equations 4.54 and 4.55 becomes:

$$\sin\beta \geq \frac{0.729 + 0.2\left(8.5 - 5\frac{\Delta h}{h}\right)\frac{L}{h}}{18.5\frac{L}{h}} \qquad (4.58)$$

Varying $\Delta h/h$ and L/h, the corresponding values of the slope angle β at limit of equilibrium are calculated and given in figure 4.18.

The application of equation 4.58 is limited up to the value of $\sin\beta \leq \sin\beta_{crit}$. Moreover, for small values of L/h the assumption of a linear decreasing excess pore water pressure along the potential sliding surface cannot be valid anymore.

So, in a quick clay mass with a slope angle β close to the critical value and in which more permeable seams are present, an increasing excess pore pressure at the upward end can easily explain the occurrence of long flake type slidings. For such large flake type slidings, the required excess pore water pressure at a given slope angle β is smaller than the corresponding value for shorter shell shaped slidings. Such conclusions are valid when the slope angle is close to its critical value β_{crit}.

4.5.4 Conclusions

Analyzing the stability problem of quick clay masses taking into account excess pore water pressures, the slope angle β seems to be limited to $\beta < 12°$ when assuming that the phreatic level coincides with the ground surface and stress conditions are changed slowly. In the case of a relatively quick change of external loading, such slope angle is reduced to $\beta < 5°30'$.

For smaller slope angles a succession of retrogressive shell shaped slidings can occur. Even with a very simple assumption of the excess pore water pressure distribution in more permeable seams within the clay mass an explanation can be found for the occurrence of long flake type slidings. They are more likely to occur when the undisturbed quick clay mass in nearly at rupture condition (β close to its critical value).

4.6 Risk of liquefaction

Soil liquefaction is a major concern for structures with or on loose sandy soils. Liquefaction can be caused by earthquakes by inducing a progressive build-up of excess pore water pressure due to cyclic shear stresses. When the pore pressure builds up to a level equal to the initial confining stress, soil loses its strength and large deformation occurs.

To evaluate the potential for soil liquefaction at a particular site, it is important to determine the soil stratigraphy and the state of the soils. While much research has been performed over the past decades to advance the techniques for assessing liquefaction potential of soils (Ishihara, 1993), simplified procedures are still the most widely used methods (e.g. Seed and Idriss, 1971; Robertson and Wride, 1998).

In the simplified approach, the cyclic stress ratio (CSR) generated at any depth of the soil deposit due to the earthquake loading can be obtained using simplified equations such as:

$$CSR = \frac{\tau_{av}}{\sigma'_{v0}} = 0.65 \left[\frac{a_{\max}}{g}\right] \left(\frac{\sigma_{v0}}{\sigma'_{v0}}\right) r_d \qquad (4.59)$$

where τ_{av} is the average cyclic shear stress, $a_{\max}$ is the maximum horizontal ground acceleration at the ground surface, g is the acceleration due to gravity,

σ_{v0} and σ'_{v0} are the total and effective vertical stresses respectively and r_d is a stress reduction factor which is a function of depth. Seed and Idriss (1971) proposed:

$$r_d = 1.0 - 0.00765 \cdot z \quad \text{if } z < 9.15m$$

$$r_d = 1.174 - 0.0267 \cdot z \quad \text{if } 9.15m < z < 23m$$

On the other hand, CSR depends on the maximum acceleration at the ground surface. If CSR exceeds the cyclic resistance ratio (CRR) at any depth of the soil deposit, soil liquefaction occurs at that depth. Here, a factor of safety against liquefaction can be defined as $FoS = CRR/CSR$. In theory, no liquefaction is expected to occur if $FS > 1$; and on the other hand, if $FS \leq 1$, liquefaction is expected.

Over the past 25 years, numerous studies have been carried out to correlate the CRR to in situ tests such as the standard penetration test (SPT), cone penetration test (CPT) and shear wave velocity measurements.

Robertson and Wride (1998) proposed a method based on CPT data. In their method, $CRR_{M=7.5}$ (related to a reference earthquake magnitude of $M = 7.5$) can be evaluated from the following simplified equations:

$$CRR_{M=7.5} = 93 \left[\frac{(q_{c1N})_{cs}}{1000} \right]^3 + 0.08 \qquad (4.60)$$

if $50 \leq (q_{c1N})_{cs} \leq 160$.

$$CRR_{M=7.5} = 0.833 \left[\frac{(q_{c1N})_{cs}}{1000} \right] + 0.05 \qquad (4.61)$$

if $(q_{c1N})_{cs} < 50$.

In the equations, $(q_{c1N})_{cs}$ is the clean sand equivalent of the stress-corrected cone tip resistance which is a function of q_c and sleeve friction f_s. In the Robertson and Wride model, a comprehensive procedure is used to determine $(q_{c1N})_{cs}$ through the use of some intermediate parameters, including the soil behavior index I_c as defined already by Robertson for identification of soils out of CPT.

Finally, a factor of safety against liquefaction for an earthquake magnitude M can be evaluated as:

$$FoS = \frac{CRR_{M=7.5}}{CSR} MSF$$

where MSF is the magnitude scaling factor to convert the $CRR_{M=7.5}$ to the equivalent CRR for the design earthquake. The recommended MSF for this CPT based method is $MSF = 174/M^{2.56}$.

4.7 Slope stability analysis of the Doeldok embankment

In the analysis of stability of the Doeldok embankment, the Strength Reduction method was mainly used by means of a finite element based program (PLAXIS). Nevertheless, the method of slices was used here as well to confirm the outcome. Soil properties for each soil type in the analysis can be found in section §6.3.4.

In the stability analysis no account was taken of the special procedure for sensitive clays since none of the soils involved classified as highly sensitive. However, the risk of liquefaction was assessed using the method proposed by Robertson and Wride (1998).

For characterizing the local seismicity at the Doeldok area, an earthquake magnitude of $M = 5.5$ was assumed and a Peak Ground Acceleration (PGA) of 0.05g was obtained from the seismic zonation map of Belgium, see figure 4.19. The embankment is located in the harbor of Antwerp, nearby point An in the figure that falls in Zone 1.

Making use of the latest data of CPT tests performed under water through the sand body of the embankment at different locations, a factor of safety could be evaluated (see Figs 4.20, 4.21, 4.22, 4.23 and 4.24). In all cases the safety factor against liquefaction did exceed 1, in fact most factors ranged from FoS = 2.5 to 6. One can conclude that liquefaction within the embankment body for an earthquake magnitude of 5.5 will not occur.

The outcome of the analysis of stability of the Doeldok embankment is given in more detail in chapter 7.

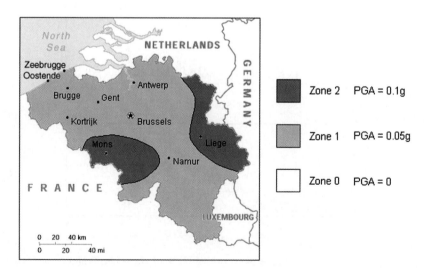

Figure 4.19. Seismic zonation of Belgium (NBN-ENV 1998-1-1: 2002 NAD-E/N/F)

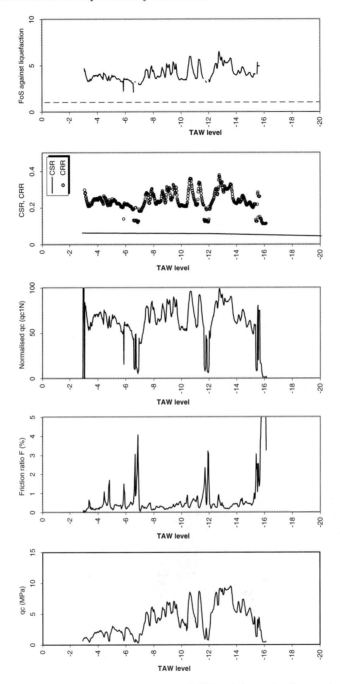

Figure 4.20. Liquefaction assessment out of CPT of the embankment body above the SSI improved zone (CPT3)

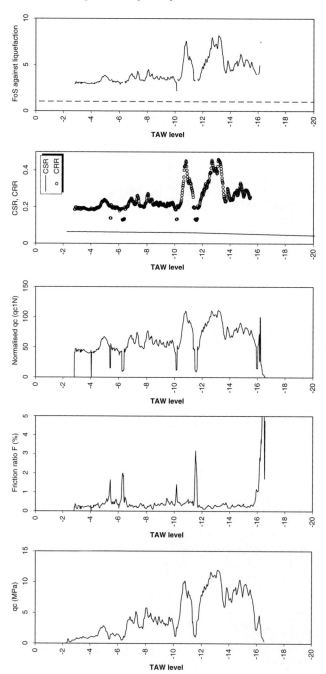

Figure 4.21. Liquefaction assessment out of CPT of the embankment body above the SSI improved zone (CPT5)

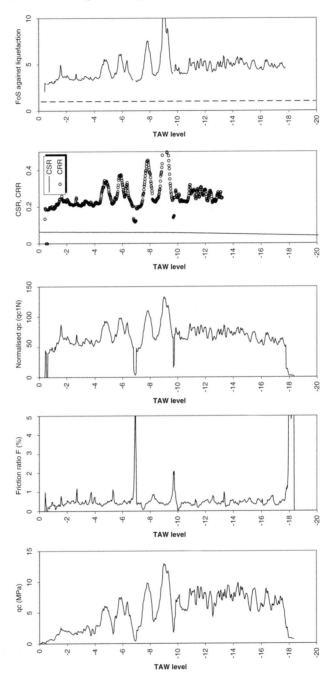

Figure 4.22. Liquefaction assessment out of CPT of the embankment body above the non-improved zone (CPT8)

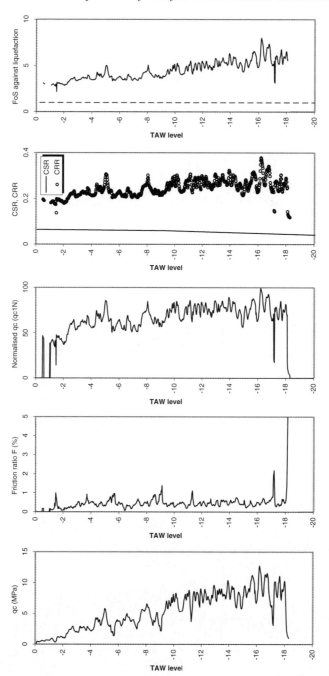

Figure 4.23. Liquefaction assessment out of CPT of the embankment body above the non-improved zone (CPT14)

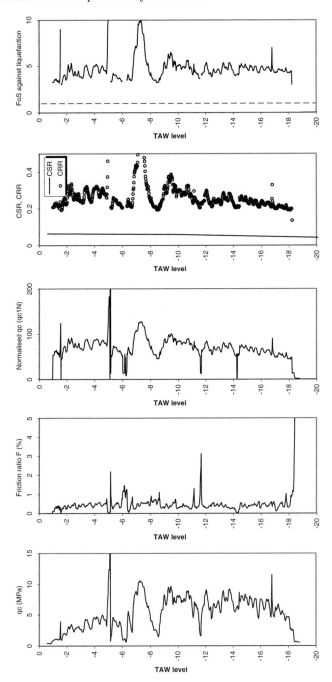

Figure 4.24. Liquefaction assessment out of CPT of the embankment body above the non-improved zone (CPT21)

5

Evaluation of consolidation

W.F. Van Impe, P.O. Van Impe & R.D. Verástegui Flores
Laboratory of Geotechnics, Ghent University, Belgium

5.1 One-dimensional consolidation theory

Saturated deposits of low permeability soils, when loaded by a surcharge (i.e. construction of an embankment on saturated clay), undergo settlements occurring over a long period of time. This phenomenon is called consolidation and it was extensively studied by Terzaghi in 1914 by means of a simple piston-spring-water model (Fig. 5.1).

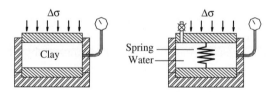

Figure 5.1. Terzaghi's piston-spring-water model

Afterward, the consolidation theory was extended and improved in many ways (e.g., Barden, 1965; Davies et al., 1965; Gibson et al., 1967; Schiffman, 1980; Gibson et al., 1981). Nowadays, two theories can be well differentiated: *infinitesimal strain* and *finite strain* theories of consolidation.

The infinitesimal strain theory assumes that the deformation of the soil mass is so small that the compressibility and the hydraulic conductivity remain constant for a given load increment. This assumption was shown to produce good results for many practical problems. However, when very soft soils are involved wherein the properties (compressibility and permeability) gradually change with consolidation progress, such assumption may lead to erroneous predictions.

Recognition of the limitations of small strain consolidation theories led to the development of large or finite strain models in which no restrictions are imposed on the deformation of the compressible media. However, the price one has to pay for a better approach is a more complex procedure that can be solved only numerically.

General and common assumptions usually accepted in the derivation are:

- The porous medium consists of incompressible pore water, incompressible mineral particles and deformable skeleton.
- The deformation consists of the rearrangement of mineral particles accompanied by a flow of pore water. The volume of soil solids remain constant.
- Skeleton is homogeneous; i.e., a single void ratio-effective stress and void ratio-hydraulic conductivity relationships govern the entire soil mass.
- The clay layer is normally consolidated.
- The water flow is one dimensional, and it is motivated by mechanical forces such as surcharge loading. Thermal, electrical or chemical potential to induce flow are not included.
- Effective stress principle applies.
- The flow of fluid through the porous skeleton is governed by the linear Darcy-Gersevanov relationship.

The next section describe each theory and solutions in more detail.

5.2 Infinitesimal strain theory

As initiated by the early work of Terzaghi on stress strain analysis of saturated soils, the one-dimensional small strain consolidation theory can be expressed in a rather easy way.

Let's consider a saturated element of porous material with characteristics fulfilling the assumptions above. In order to obtain the equation governing the consolidation, we start from the continuity requirement for the solid and liquid interaction. The continuity equation for the solid phase requires:

$$\frac{\partial \left[(1-n)\rho_s v_s\right]}{\partial z} + \frac{\partial \left[(1-n)\rho_s\right]}{\partial t} = 0 \tag{5.1}$$

In the same way, the continuity equation for the fluid requires:

$$\frac{\partial \left[n\rho_w v_w\right]}{\partial z} + \frac{\partial \left[n\rho_w\right]}{\partial t} = 0 \tag{5.2}$$

where v_s and v_w are the absolute velocities of the solid and liquid phase respectively in an Eulerian coordinate system; ρ_s and ρ_w are the unit weight of the solid and fluid phase; n is the porosity of the element, z is the vertical coordinate and t is the time.

Combining equations 5.1 and 5.2, the continuity condition for the mixture will be:

$$\frac{\partial \left[(1-n)v_s + nv_w\right]}{\partial z} = 0 \tag{5.3}$$

Considering that Darcy's law still governs the flow of water through the soil skeleton (i.e. accepting all boundary conditions limited to a Darcy type of flow), the equilibrium equation of water can be expressed as:

$$\frac{\partial u}{\partial z} + n\frac{\gamma_w}{k}(v_w - v_s) = 0 \qquad (5.4)$$

where u is the excess pore pressure and k is the hydraulic conductivity.

Differentiation of 5.4 with respect to z leads to:

$$\frac{\partial^2 u}{\partial z^2} - \frac{\gamma_w}{k}\frac{\partial [n(v_w - v_s)]}{\partial z} - \frac{\gamma_w}{k^2}\frac{\partial k}{\partial z}n(v_w - v_s) = 0 \qquad (5.5)$$

and combining with equation 5.3, it gives:

$$\frac{\partial^2 u}{\partial z^2} - \frac{\gamma_w}{k}\frac{\partial v_s}{\partial z} + \frac{1}{k}\frac{\partial k}{\partial z}\frac{\partial u}{\partial z} = 0 \qquad (5.6)$$

The partial derivative $\partial v_s/\partial z$ can be expressed in terms of settlement rate, in order to introduce the constitutive relationship for the soil skeleton, therefore, the following holds true:

$$\frac{\partial v_s}{\partial z} = -\frac{\partial \epsilon_1}{\partial t} = -m_v\frac{\partial \sigma'_v}{\partial t} \qquad (5.7)$$

and the previous equation (5.6) can be written as:

$$\frac{\partial^2 u}{\partial z^2} + \frac{\gamma_w}{k}m_v\frac{\partial \sigma'_v}{\partial t} + \frac{1}{k}\frac{\partial k}{\partial z}\frac{\partial u}{\partial z} = 0 \qquad (5.8)$$

The change of vertical effective stress with time can be expressed in terms of change of pressure applied at the surface and change of excess pore water pressure:

$$\frac{\partial \sigma'_v}{\partial t} = \frac{\partial q}{\partial t} - \frac{\partial u}{\partial t} \qquad (5.9)$$

and by substituting 5.9 in 5.8, the general equation of one-dimensional consolidation can be expressed in terms of excess pore pressure:

$$\frac{\partial^2 u}{\partial z^2} + m_v\frac{\gamma_w}{k}\left(\frac{\partial q}{\partial t} - \frac{\partial u}{\partial t}\right) + \frac{1}{k}\frac{\partial k}{\partial z}\frac{\partial u}{\partial z} = 0 \qquad (5.10)$$

Now, If we assume that the hydraulic conductivity remains constant, then equation 5.10 reduces to:

$$c_v\frac{\partial^2 u}{\partial z^2} = \frac{\partial u}{\partial t} - \frac{\partial q}{\partial t} \qquad (5.11)$$

where $c_v = k/(m_v\gamma_w)$, is the coefficient of consolidation and the term $\partial q/\partial t$ describes the change of surcharge load with time (variable load).

Then, equation 5.11 allows modeling of consolidation due to variable loading (i.e. staged construction) when changes of hydraulic properties during loading are neglectable.

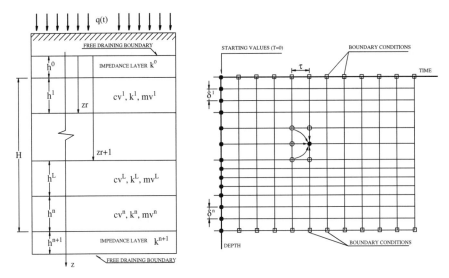

Figure 5.2. Multi-layered system and finite difference solution scheme

5.2.1 Numerical solution

The one-dimensional consolidation is governed by the parabolic partial differential equation 5.11. The problem is classified as one of initial value since we know in advance the excess of pore water pressure at a time $t_0 = 0$ and we want to estimate a new value at a time $t = t_0 + \Delta t$. The problem can also be extended to allow several layers of clay with different properties, a multi layered system (Fig. 5.2).

It is assumed that the soil profile consists of n contiguous layers. The arbitrary layer is indexed L with thickness h^L. The soil properties of the Lth layer are the coefficient of consolidation c_v^L, the compressibility m_v^L, and the coefficient of permeability k^L. The compressible stratum is the system of n compressible layers and has a total thickness H.

The analysis that follows describes the solution proposed by Schiffman and Arya (1977).

Indexes which refer to a layer are written as superscripts. The superscript L refers to an arbitrary layer. Thus, all soil properties are superscripted. The space coordinate z is a global coordinate and has its origin at the surface $z = 0$ (Fig. 5.2). All indexes that depend on the global space coordinate z are written as subscripts. Similarly, all indexes that are dependent of time are written as subscripts.

As a general rule the superscript L refers to a layer number; the subscript i refers to a z coordinate point and the subscript j refers to a value at a particular time. The value of the superscript L runs from 1 to n. The value of i runs from 0 at the surface ($z = 0$) to a value p at the lower boundary ($z = H$).

The subscripted space index at the layer interface is r. The subscripted time index j runs from 0 at $t = 0$ in an arithmetic progression $(0, 1, 2, 3)$.

The consolidation phenomena of a single layer is governed by equation 5.11 alone. However, in a multi-layered system there are n similar equations which must be solved in order to determine the excess pore water pressure at any point in space and time. Equation 5.11 can be rewritten as:

$$c_v^L \frac{\partial^2 u^L}{\partial z^2} = \frac{\partial u^L}{\partial t} - \frac{dq}{dt} \qquad L = 1, 2, ..., n \qquad (5.12)$$

The three types of time-independent boundary conditions that can apply to the stratum boundaries $z = 0$ and $z = H$ can be expressed in general form as:

$$a^1 \frac{\partial u^1}{\partial z}(0, t) - b^1 u^1(0, t) = -c^1 \qquad (5.13)$$

$$a^n \frac{\partial u^n}{\partial z}(H, t) - b^n u^n(H, t) = -c^n \qquad (5.14)$$

where the coefficients a^1, b^1, c^1, a^n, b^n and c^n take on specific values for specific conditions (i.e. free draining, impervious and partial drainage or impeded). Table 5.1 presents the particular values of these coefficients.

Table 5.1. Boundary conditions

Boundary condition	Upper boundary a^1	b^1	c^1	Lower boundary a^n	b^n	c^n
Free draining	0	-1	0	0	1	0
Impervious	1	0	0	1	0	0
Impeded	h^1	$\lambda^1 = \frac{k^0 h^1}{k^i h^0}$	0	h^n	$\lambda^n = \frac{k^{n+1} h^n}{k^n h^{n+1}}$	0

In addition to the boundary conditions for the compressible stratum, it is assumed that there is full continuity between clay layers. This assumption requires that the excess pore water pressure and flow velocities in adjacent layers are equal at the common layer interfaces. It gives:

$$u^L(z_r, t) = u^{L+1}(z_r, t) \quad \text{and} \quad k^L \frac{\partial u^L}{\partial z}(z_r, t) = k^{L+1} \frac{\partial u^{L+1}}{\partial z}(z_r, t) \qquad (5.15)$$

where, as shown in figure 5.2, the distance z_r is the distance from the surface to the layer interface separating the L^{st} and the $(L+1)^{\text{st}}$ layer. So far, all the formulation has been described.

A solution can be applied at this point making use of either finite difference methods or finite elements. The finite difference procedure was chosen here. The finite difference method makes use of a discretization of the space and time and therefore replaces the continuous derivatives by the ratio of changes in the variable over a small but finite increments.

A program, SSCON-FD (Small Strain CONsolidation – Finite Difference solution), has been developed here on the basis of the previous procedure. The program was written in Maple. As explained before, it allows the infinitesimal strain consolidation analysis of a multilayered soil deposit. In the model each layer may have different consolidation parameters and the boundary conditions allow not only fully drained or fully closed conditions but also partial drainage by introducing hydraulic conductivity values of impedance layers (see Fig. 5.2).

5.2.2 Applications of SSCON-FD

In this section the range of applications of the program are described, with special attention to the analysis of staged loading. For this purpose, 2 fictitious problems have been addressed and numerical solutions obtained.

5.2.2.1 Problem 1

Let's assume that a sand embankment is constructed in stages over a saturated clay layer. The clay layer has a thickness of 5 m, $c_v = 3\,\text{m}^2/\text{year}$ and $k = 0.02\,\text{m/year}$, it rests over a silty layer with a thickness of 1 meter with $k = 0.03\,\text{m/year}$. The silty layer rests over a deep sand with high permeability. The first load stage (80 kPa) was placed uniformly over a period of 5 months. The work was stopped for 1 year and finally, the second load stage was applied with the same magnitude and at the same rate.

It is required to study the consolidation of the clay layer:

5.2.2.2 Solution

SSCON-FD was used to solve the problem. Free draining conditions were adopted for the upper boundary and impeded conditions for the lower since there might exist downwards flow. Results are shown in figures 5.3 and 5.4.

Figure 5.3 illustrates the development of effective stress with time (at the midheight of the layer) and figure 5.4 describes the excess pore pressure in time and space.

5.2.2.3 Problem 2

An instantaneous load of 100 kPa has been applied at the top of a 4-meter thick soil deposit that rests over a sound rock with very low permeability. The soil deposit consists of two different clays layers with the same thickness. The consolidation properties of the upper clay are: $c_v = 1.5\,\text{m}^2/\text{year}$ and $k = 0.4\,\text{m/year}$. The consolidation properties of the lower clay layer are: $c_v = 0.2\,\text{m}^2/\text{year}$ and $k = 0.1\,\text{m/year}$.

It is required to study the consolidation of the layered system.

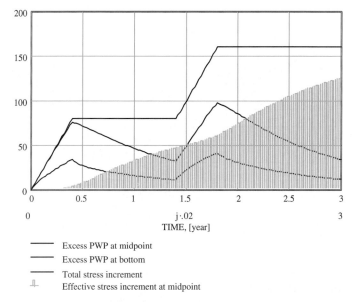

Figure 5.3. Consolidation analysis of staged loading (SSCON-FD)

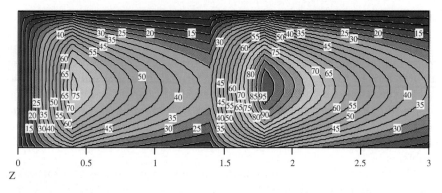

Figure 5.4. Contour lines of excess pore water pressure (SSCON-FD) as a function of space (vert. axis) and time (horiz. axis)

5.2.2.4 Solution

Again, SSCON-FD was used to solve the problem. Free draining conditions were adopted at the top boundary and impermeable at the bottom. The results are shown in figure 5.5. From the excess pore water pressure profiles, numerically predicted up to 3 years, one can clearly note that the one-dimensional consolidation of the upper layer goes faster than in case of the lower layer due to the considerable difference in consolidation parameters.

Since SSCON-FD allows the modeling of multilayered systems, thick layers of clay showing different parameters with depth could be modeled as such,

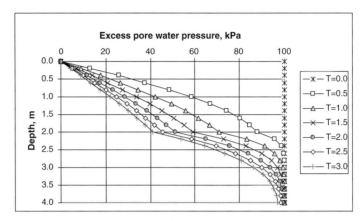

Figure 5.5. Excess pore water pressure profiles (SSCON-FD)

in this way, avoiding having to rely on an average consolidation parameter for the whole strata. Such practice usually leads to erroneous predictions as demonstrated by Pyrah (1996).

5.3 Finite strain theory

General three-dimensional consolidation approaches have been discussed already in literature starting from the important work of Biot (1941); however, all of it related to small strain levels of deformations of the soil skeleton.

The most general theory of a one-dimensional type of consolidation is that proposed by Gibson et al. (1967). This analysis overcomes the limitations that the classical, small strain, theory entails; but at the same time the problem becomes so complex that only numerical solutions can be obtained for practical problems.

The process of finite strain one-dimensional consolidation of a saturated porous medium is governed by:

$$\frac{\partial}{\partial z}\left[g(e)\frac{\partial e}{\partial z}\right] - b(e)\frac{\partial e}{\partial z} + \frac{\partial e}{\partial t} = 0 \qquad (5.16)$$

where

$$g(e) = -\frac{k(e)}{\gamma_w(1+e)}\frac{d\sigma'}{de}$$

$$b(e) = \left(\frac{\gamma_s}{\gamma_w} - 1\right)\frac{d}{de}\frac{k(e)}{1+e}$$

in which e is the void ratio, γ_s and γ_w are the solid and fluid weights per unit of their own volume, respectively, and z is a reduced coordinate encompassing a volume of solids in a volume of unit cross sectional area lying between the datum plane and the Lagrangian coordinate point (Gibson, 1967).

The function $g(e)$ plays the role of consolidation coefficient and $b(e)$ introduces the effect of gravity. If the gravity effect is neglected [i.e. $b(e) = 0$] and $g(e)$ is assumed to remain constant during the process, then equation 5.16 would simplify to the classical theory (i.e. Terzaghi's).

Equation 5.16, with appropriate boundary and initial conditions and constitutive properties, provides the governing relationship from which a solution can be developed. The required constitutive properties are: the relationship between void ratio and effective stress, and the relationship between coefficient of permeability and void ratio.

It is noted that the governing equation, while unrestricted as to the magnitude of strain and non linearity of constitutive relationships, it is based upon the premise of homogeneity and monotonic behavior; load-unload-reload cycles are not permissible.

Many researchers have attempted to solve the problem (Cargill, 1984; Fox and Berles, 1997; Van Impe P.O., 1999). The work of Van Impe P.O. (1999) extended the formulation introducing the sedimentation phase of a soil deposit during its skeleton formation. Within this framework, two programs were elaborated: FISCC (Van Impe P.O., 1999) and CBFISCC. Both analyze the one dimensional consolidation of a single clay layer subjected to the action of an instantaneous constant load introducing different numerical methods for the solution of the governing differential equation.

- CBFISCC is a numerical solution of the partial non-linear differential equation governing the phenomena. The equation is transformed to a system of ordinary differential equations and the system is solved using an appropriate algorithm.
- FISCC is piecewise linear model. It encompasses an iterative procedure in which each step is linearly evaluated, but it becomes non linear in the overall solution. To apply this technique it is essential to use sufficiently small time steps; therefore, it obviously requires more time for computation.

The input data of such programs usually consist of:

- initial thickness of the compressible layer
- specific gravity of solids and water
- initial void ratio of the compressible layer
- boundary drainage conditions (open or close)
- total stress increment at the surface
- constitutive relationships of the soil (e.g. $K = K(\sigma'_v)$ and $e = e(\sigma'_v)$)

The program output consist of:

- settlement versus time
- void ratio profiles at several time steps
- pore water pressure profiles at several time steps

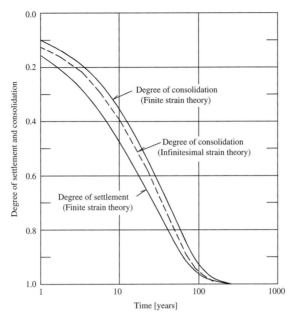

Figure 5.6. Degrees of settlement and consolidation (pore pressure dissipation) (Schiffman, 1994)

5.4 Infinitesimal strain versus finite strain theory

As demonstrated in the previous section the conventional one-dimensional consolidation theory is nothing but a special case of the more general theory formulated by Gibson et al. (1967).

Figure 5.6 shows the results of a comparative study performed for a 10 m thick, normally consolidated St. Herblain clay layer subjected to a surcharge of 200 kPa (Schiffman, 1994).

In finite strain consolidation analysis, unlike in infinitesimal theories, there are two measures of the progress of consolidation: the degree of settlement (ratio of settlement at a time t to the final settlement) and the degree of pore water pressure dissipation.

As shown in figure 5.6, the conventional (infinitesimal) theory underpredicts the rate of settlement and overpredicts the rate of pore water pressure dissipation. Clearly, the use of finite strain theory would provide a better (safer) estimates of the shear strength gain of a consolidating clay.

5.5 Consolidation at the Doeldok site

As far as the actual case of the embankment foundation soil is concerned, a simulation was performed to evaluate both, the progress of settlements and

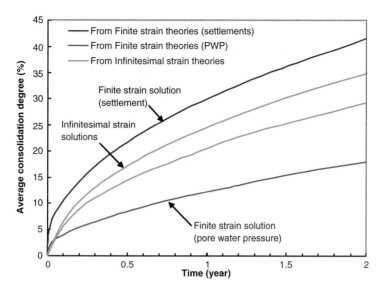

Figure 5.7. Consolidation analysis of the foundation soil evaluated with large and small strain theories

pore water pressure dissipation. In this simulation, both theories of consolidation (large strain and small strain) were applied implementing the constitutive equations ($K = K1(\sigma'_v)$ and $K = K2(e)$) for the soft dredged material obtained from several tests (see section §6.3.1.3). In this simulation, a single load increment equal to the weight of the whole embankment has been studied. The results are illustrated in figure 5.7 which confirm that the excess pore water pressure dissipation and the settlement will take place at different rates. The output of small strain analysis in figure 5.7 is showed as a range because there is a range of consolidation coefficients that can be chosen out of the constitutive equations.

Clearly, the pore water pressure is expected to dissipate quite slowly (after 2 years only 15% of the total excess will dissipate) while significant settlements are expected to be observed in the same time period (after 2 years already 40% of the total settlement will occur). Moreover, it can be concluded that small strain theories overestimate the degree of consolidation from excess pore water pressure but they underestimate the degree of consolidation from settlements.

As it is discussed in chapter 7, the presence of the soft soil as foundation soil for the underwater embankment required a controlled rate of construction in the framework of a staged construction. Staged construction allows the foundation soil to partially dissipate its excess pore water pressure and relies on a strength increase that depends on the consolidation degree. This means that measuring or estimating the consolidation degree of the foundation soil during construction is of utmost importance. Therefore, the foundation soil was provided with a number of pore water pressure transducers at several location and at 3 different levels within the layer. Moreover, settlement profiles

68   Evaluation of consolidation

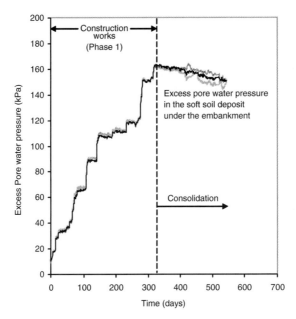

Figure 5.8. Excess pore water pressure at one location in the foundation soil

under the embankment were also regularly measured by means of a clever and simple system that makes use of a plastic tube that is laid under the embankment and that is filled with water. Then, settlement can be derived out of water pressure (head) measurements inside the tube by means of a probe that is pulled along the plastic tube. Those 2 instrumentation means allowed for a controlled construction.

Figures 5.8 and 5.9 illustrate the progress of excess pore water pressure and settlements under the embankment load during construction up to the current situation. A more detailed explanation of the construction procedure can be found in chapter 9, but in a nutshell, the first phase of the dam was built in stages allowing for some time in between. At the end of Phase 1 an even longer period was introduced to allow the soft soil to consolidate.

The results of the monitoring of pore water pressures and settlements do show indeed that there has been a very slow dissipation of pore water pressure and a much faster progress of settlements.

A more elaborated analysis of the consolidation progress at the Doeldok site under the current loading situation was attempted here and results were compared to measurements (see Figs 5.10 and 5.11). As expected from the previous discussions it can be observed that the large strain consolidation prediction match the measurements more accurately than small strain predictions. The small deviation observed may just be consequence of the natural inhomogeneity of the soft foundation soil.

Consolidation at the Doeldok site 69

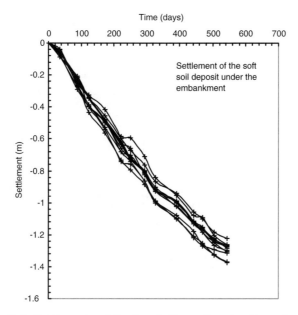

Figure 5.9. Settlements of the foundation soil under the embankment

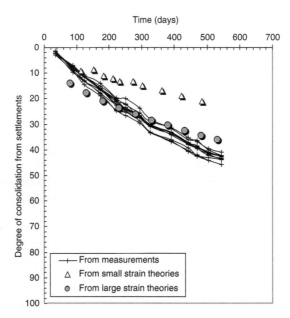

Figure 5.10. Degrees of consolidation of the foundation soil studied here out of settlements

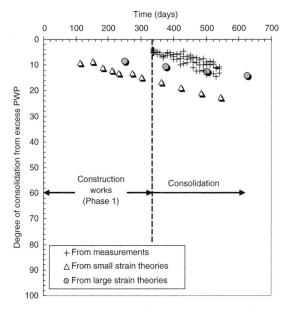

Figure 5.11. Degrees of consolidation of the foundation soil studied here out of pore water pressure

5.6 Conclusions

The finite strain theory constitutes outspokenly a better approach to the phenomena of consolidation. It was shown that the use of infinitesimal strain consolidation theories may be underconservative for staged construction analysis.

Nevertheless, at the very first stage of design, the classical theory may still provide the basis for further refinement of calculations. Clearly, many geotechnical engineers are more familiarized with consolidation parameters for the classical (infinitesimal strain) theory than the finite strain theory. As a rule, the softer the soil the larger the error introduced in the analysis by not taking into account changes of compressibility and hydraulic conductivity during consolidation.

6

Geotechnical characterization of the site

W.F. Van Impe & R.D. Verástegui Flores
Laboratory of Geotechnics, Ghent University, Belgium

J. Van Mieghem & A. Baertsoen
Ministry of Flanders, Belgium

6.1 Overview

The present chapter illustrates the geotechnical characterization of the soils present in the site. Also some estimations are proposed of the expected properties of the sand embankment body.

The site of investigation is a dock located in the harbor of Antwerp (Belgium). The water depth to the sediments level is about 20 m in the area.

6.2 Soil profile and characterization

In the framework of a geotechnical investigation of the foundation soil at the site, a number of field tests such as CPT and field vane tests were carried out focusing on the characterization of the soft material. Figure 6.1 illustrates the location of the tests in the area. Moreover, several borings were performed to collect soil for laboratory testing. All field tests were carried out from a jack-up platform.

Figure 6.2 illustrates a typical CPT profile. Clearly the soft material (with $q_c < 0.3$ MPa) extends for about 8 m. Underneath, a relatively thin (thickness of 2 m) sand layer can be found resting on a deep highly overconsolidated clay from the Tertiary (Boom clay).

Out of the field vane testing it was possible to estimate not only the undrained shear strength at several depths but also to confirm the normal consolidation state of this young deposit. Figure 6.3 illustrates the measured undrained shear strength c_u plotted against the estimated vertical effective stress. With the exception of some points, the general trend seems to be linear increasing with depth (with $c_u \approx 0.3\sigma'_v$) as it would be expected in young normally consolidated soils. Points deviating from the trend in the upper part of the layer suggest the presence of a lightly overconsolidated crust; on the other hand, points deviating from the trend in the lower layer may actually belong to a transition zone of soft material and sand.

72 Geotechnical characterization of the site

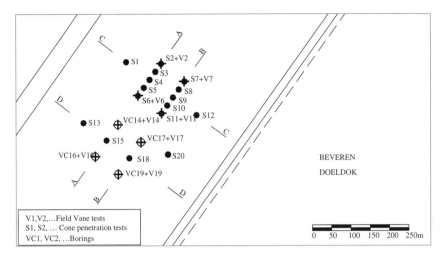

Figure 6.1. Location of field tests

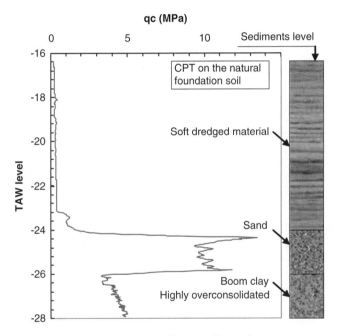

Figure 6.2. Typical CPT profile in the area

6.3 Selection of parameters for design

The design of the embankment involves different materials such as: sand (for the embankment itself), Boom clay and dredged sludge (with and without improvement). Each material and its properties are described below.

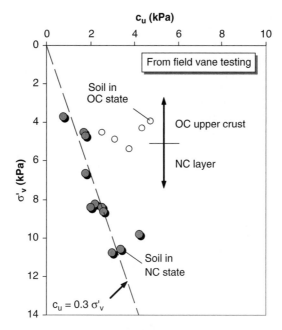

Figure 6.3. Field vane test results

6.3.1 Dredged material

The soft dredged material is actually the result of years of sedimentation and self-weight consolidation of dredged sediments from waterways within the harbor of Antwerp. Although there were attempts in the past to improve this material by vacuum consolidation, its consistency remained soft.

6.3.1.1 Physical properties

Table 6.1 summarizes some averaged physical properties of the material in its natural state. The natural water content of the soil is of the order of 115%, the plasticity index of the order 77 and the organic content of about 6%. pH measurements of the pore water give a value close to 7.

In order to get a closer look of this material, a specimen of untreated dredged material was analyzed on the Scanning Electron Microscope (SEM).

The scanning electron microscope is a type of electron microscope capable of producing high resolution images of a sample surface at magnification levels that could go up to molecular levels. These images have a characteristic 3D appearance and are useful for judging the microstructure of a sample.

The working principle of SEM is simple, electrons are shot to a sample and as a result of their interaction (electrons and matter) electrons from the sample are released. Such released electrons provide information of the sample,

Table 6.1. Physical properties of the dredged material

Index	Value
Liquid limit	124.4
Plastic limit	46.7
Plastic index	77.7
Natural water content, %	115.0
Organic content, %	6.0
Natural lime content, %	13.9
Sand fraction, %	10.4
Wet density, g/cm^3	1.4
pH of the pore water	7.2

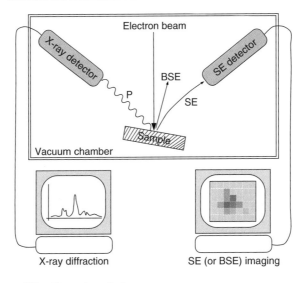

SE: Secondary electrons
BSE: Back-scattered electrons
P: Photons

Figure 6.4. Scanning electron microscopy working principle

therefore they are detected and analyzed (Fig. 6.4). The primary electron beam is produced by an electron gun and accelerated toward the sample. The sample must be conductive so that the energy of the inciding electrons can be diverted from the sample, otherwise, the electron beam could burn and destroy the sample before any analysis can be done. That is not an issue when analyzing steel samples, but in the case of soils (non conductive) the samples should be coated with a fine (nano scale) layer of a conductive element such as gold. Moreover, samples are tested dry in a vacuum chamber to minimize any interaction of the inciding and emitted electrons (from the sample) with air particles. However, newly developed SEM allow testing samples with some amount of water and coating is no longer mandatory.

Selection of parameters for design   75

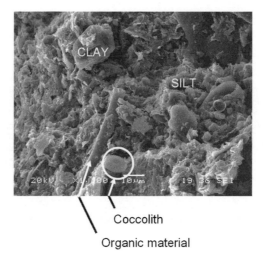

Figure 6.5. Scanning electron microscopy analysis of natural dredged material

The electron beam is deflected by scanning coils so that it can raster an area of the sample surface. As the primary electrons strike the surface they are elastically and inelastically scattered by atoms in the sample. Through these scattering events, the primary electron beam effectively spreads and fills a teardrop-shaped volume, known as the interaction volume, extending about less than 100 nm to 5 m depths into the surface. Interactions in this region lead to the subsequent emission of electrons and photons (e.g. X-rays) which are then detected to produce an image. Two main types of emitted electrons can be identified: backscattered electron (BSE) which are high energy electrons product of elastic interaction of the electron beam with the sample and secondary electrons (SE) which are low energy electrons from inelastic interactions.

X-rays, usually detected in a SEM, allow the study of the chemistry of a particular point in the surface of the sample by identification techniques based on energy dispersive X-ray spectroscopy or wavelength dispersive X-ray spectroscopy.

SE electrons (more commonly used for SEM imaging) are detected by a scintillator-photomultiplier device and the resulting signal is rendered into a two-dimensional intensity distribution that can be viewed and saved as a digital image for each point on the sample surface that the electron beam rasters. Finally, a sharp image of the surface can be generated.

The specimen analyzed under the SEM was prepared to simulate the natural in situ conditions of the dredged material with very light hand compaction. Figure 6.5 does illustrate a SEM picture of the specimen at an amplifying factor of 1100. Clay particles (platy shaped) can be identified by their somewhat more bright color in the picture. They seem to be uniformly spread and are interacting with the silt and sand particles (in edgy or rounded shape and

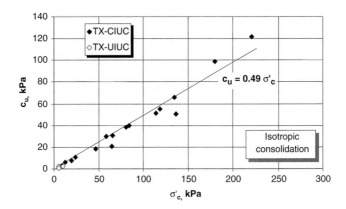

Figure 6.6. Undrained strength ratio from laboratory testing

much darker in color). It is also possible to find organic matter and biological remains showing a very regular micro morphology, typical in marine soils.

6.3.1.2 Mechanical properties

As illustrated in figure 6.3, the undrained strength of the dredged material is quite low. It was found that c_u ranges from 2 to 4 kPa. For the design phase, an initial value of $c_u = 3$ kPa has been assumed as the most representative.

Furthermore, laboratory tests have been focused on the evaluation of a undrained strength ratio ($S_0 = c_u/\sigma'_v$). From a series of triaxial CU tests on isotropically consolidated specimens (Fig. 6.6), a value of $S_0 \approx 0.49$ was found. This confirmed also the fact that this soil has a normalized behavior.

However, it is known from literature (Kousoftas, 1981) that isotropic consolidation may overestimate the actual S_0. In fact, the undrained strength ratio estimated out of field vane tests is only of the order of 0.3. Then a $S_0 = 0.3$ was chosen for the design.

As for the stiffness of the material an estimation of the drained Young's modulus was attempted by means of the graph in figure 6.7.

Figure 6.7 illustrates a chart to correlate the elastic Young's modulus of natural clay from the undrained strength, the overconsolidation ratio and the plasticity index. We should look at the OCR = 1 level corresponding to $PI > 50$, therefore, we obtain $E'_y/c_u \approx 150$. This conservative ratio has been assumed for natural dredged material and also as a reference for the dredged improved material.

In conclusion, the parameters adopted for *natural* dredged sludge are: $c_u \approx 3$ kPa, $c' = 0$, $\phi' = 18°$, $\gamma = 12.8$ kN/m^3, and a drained Young's modulus of $E'_{50} = 0.5$ MPa.

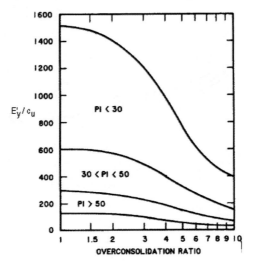

Figure 6.7. Empirical correlation between E'_y/c_u, OCR and PI (USACE, 1990)

6.3.1.3 Consolidation properties

The consolidation behavior of the dredged material has been assessed by means of Constant Rate of Strain (CRS) tests, hydraulic conductivity tests and oedometer tests. Figures 5 and 6 summarize the results of all tests performed. A best fitting curve have been drawn on each graph; in this way the following constitutive equations necessary for large strain consolidation analysis have been evaluated:

$$k = 6.0 \cdot 10^{-8} \sigma_v'^{-1.18} \tag{6.1}$$

$$k = 6.0 \cdot 10^{-12} e^{5.52} \tag{6.2}$$

where the effective stress (σ_v') and the permeability (k) are expressed in kPa and m/s respectively.

6.3.1.4 Parameters of deep mixing improved zones

As far as strength of DM improved sludge is concerned, one may expect to obtain a cone resistance varying from $q_c = 2$ to 3.5 MPa (Van Impe W.F., 2000) for a cement content of 5% to 15%. Therefore, with $c_u \approx q_c/15$ (Nuyens et al., 1995), a conservative mean undrained strength for two levels of improvement (full and partial) would be about 100 kPa and about 50 kPa respectively. These are not prediction but design values (minimum values) that will have to be realized in the field.

Then, parameters adopted for *partially improved* sludge are as follows: $c_u \approx 50$ kPa, $c' = 0$, $\phi' = 25°$, $\gamma = 13.5$ kN/m^3, and $E'_{50} = 7.5$ MPa.

78 Geotechnical characterization of the site

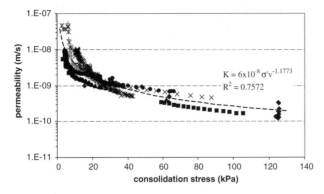

Figure 6.8. Hydraulic conductivity vs. vertical effective stress

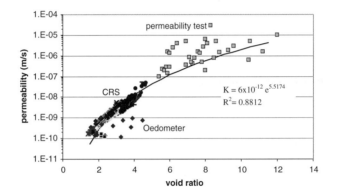

Figure 6.9. Hydraulic conductivity vs. void ratio

6.3.2 Boom clay

The bottom layer on the soil profile consists of Boom clay. This material has also been extensively studied in a number of papers (i.e. De Beer, 1967; Bouazza et al., 1996).

De Beer (1967) studied the shear characteristics of the Boom clay for a project of a tunnel under the Scheldt river in Antwerp, nearby the current site of investigation.

According to a geological study carried out at that site, the Boom clay was covered by about 40 m of Antwerpian sand at the beginning of the continental Pleistocene erosion. This load acted on the Boom clay for 5 to 7 million years and the unloading due to erosion started 500000 years ago. For that reason Boom clay exhibits a brittle stress strain behavior, typical of overconsolidated clay.

For the determination of undrained shear strength a large number of unconfined compression tests and undrained triaxial tests were performed

(De Beer, 1967) on samples extracted to depths of about 50 m. The following relationship was deduced:

$$c_u \, (\text{kPa}) = 75 + 3.5 D \, (\text{m}) \qquad (6.3)$$

where c_u is the undrained shear strength expressed in (kPa) and D is the depth expressed in (m).

Simultaneously, in order to determine the drained shear strength parameters a large number of consolidated undrained triaxial tests were carried out. Due to considerable scatter, a range of friction angles was obtained, varying from 17° to 24° and from 15° to 19° estimated with the peak and residual strength respectively.

Bouazza et al. (1996) studied mechanical properties of reconstituted Boom clay. The authors published stress-strain curves of shear tests from which the undrained Young's modulus has been estimated as round $E_u = 50$ MPa and the drained Young's modulus was estimated as $E' = 40$ MPa.

Small strain stiffness measurements (Haegeman, 1999) showed that the Boom clay, under low levels of mean stress, has a E'_{max} (small strain Young's modulus) in the order of 80 MPa. Since the strain level for the problem in consideration is certainly higher, the choice of drained Young's modulus $E'_{50} = 40$ MPa seems to be reasonable.

Moreover, Bouazza et al. (1996) provide information about the intrinsic properties (critical state concept) of the Boom clay. They found that $\phi'_{cs} = 18.5°$.

Then, the following parameters have been adopted for Boom clay: $c_u \approx 100$ kPa, $c' = 0$, $\phi' = 19°$, $\gamma = 19$ kN/m³, and a drained Young's modulus of $E'_{50} = 40$ MPa.

6.3.3 Sand

As far as the embankment sand parameters are concerned, there are already some experiences with sand dumping in harbor areas. For example, the construction of the breakwater at the new outer harbor of Zeebrugge in Belgium. A number of publications (De Wolf et al., 1983; Van Impe W.F., 1985; Van Impe W.F., 1989) point up properties of this material.

The sand employed in that project consisted of rather coarse quartz particles with shells. The sand was used as replacing material in order to improve the foundation of the breakwater. The sand was dumped with hopper suction dredgers.

Here, it was assumed that a similar construction procedure would go on, then, we can expect similar parameters. According to results obtained following a quality control, the cone resistance in the sand after dumping varied from 6 to 10 MPa corresponding to rather dense sand. Shear angles varying in a range from 28° to 35° were reported.

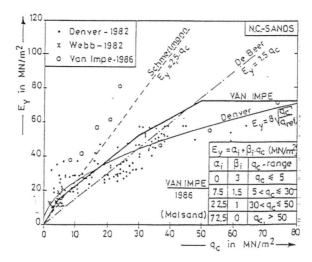

Figure 6.10. Young modulus vs. cone resistance for normally consolidated sands (Van Impe W.F., 1986)

Moreover, we can also estimate deformation parameters from well established correlations. Figure 6.10 allows correlating the Young's modulus from the cone pressure on NC sands.

In this way, the following parameters have been adopted here: $c' = 0$; $\phi' = 32°$, $\gamma = 20\,\text{kN/m}^3$, $\gamma_d = 17\,\text{kN/m}^3$, and the drained Young's modulus $E'_y \approx 15\,\text{MPa}$.

6.3.4 Summary of soil properties

Table 6.2 summarizes parameters adopted for design for each soil type.

Table 6.2. Soil parameters

Soil type	c' (kPa)	ϕ' (°)	c_u (kPa)	γ (kN/m³)	γ_d (kN/m³)	E'_{50} (MPa)	ν'
Natural dredged material	0	18	5	14	0.75	0.33	
Improved dredged material	0	18	50	16	7.5	0.33	
Sand	0	32	–	20	17	15	0.33
Boom clay	0	19	100	19	18	40	0.35

7

Design of underwater embankment of soft soil

W.F. Van Impe & R.D. Verástegui Flores
Laboratory of Geotechnics, Ghent University, Belgium

H. Cecat & C. Boone
TECHNUM, Belgium

7.1 Overview

The present chapter illustrates the design of the underwater sand embankment on soft soil (dredged material). The water depth to the sediments level is about 20 m. As detailed in the previous chapter, the soil profile at the site consists of an upper soft layer overlying sand and a deep tertiary clay layer.

Several alternatives have been evaluated. More details one these alternatives can be found elsewhere (Verástegui, 2001). Figure 7.1 illustrates a scheme of each of the options already studied:

- Alternative A: The embankment implements full improvement at the toe of the slope on the open river side while no improvement is specified for the toe at the dry dock side.
- Alternative B: The embankment comprises full improvement at the open river side and partial improvement at the dry dock side.
- Alternative C: In this configuration, the geometry slightly changes, partial improvement of the soft material is considered on both toes.
- Alternative D: The embankment has the same configuration as Alternative C except for the absence of improvement at the toe on the dry dock side.
- Alternative E: All the soft material is removed and the sand embankment is directly founded on the Boom clay (Fig. 7.1). Technically this could be qualified as a good solution, however, one must take into account that the removal and new disposal of huge volumes of dredged material may be a serious environmental issue.

The characteristics of the problem and the conditions of the foundation soil demanded improvement of the foundation (by deep mixing techniques) soil on the one hand and construction in stages on the other hand in order to prevent early instability. Moreover, restrictions imposed on the total time of construction led to adopt extra reinforcement elements such as geotextiles in the embankment body.

82   Design of underwater embankment of soft soil

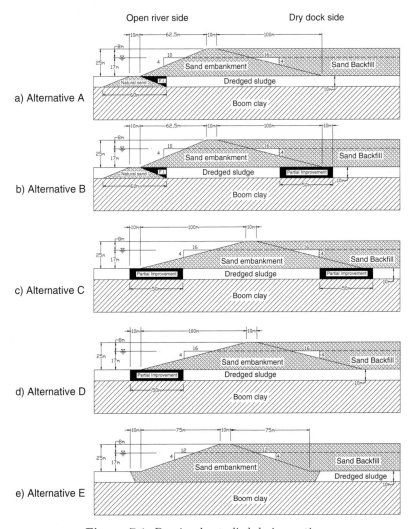

Figure 7.1. Previously studied design options

On the basis of the previous design alternatives, a more refined design has been proposed (Fig. 7.2). A partial level of improvement is considered here at the toes of the embankment.

Results of the design have shown that the construction can be safely carried out in two main phases with a extended waiting time delay in between.

7.2 Geometry of the embankment

The general profile of the embankment and soil conditions are illustrated in figure 7.2. As already described, the soil profile consists of about 8 m of soft

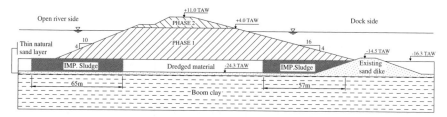

Figure 7.2. Embankment geometry final design option

material, 2 m of sand and a deep highly overconsolidated clay (Boom clay). The water depth to the sediments level is about 20 m.

The soft material was once subjected to vacuum consolidation to accelerate the self-weight consolidation. That means that there exist drains installed in the soft material that could help at accelerating the pore water pressure dissipation. However, no real evidence exits to support such positive effect of drains since there is no evidence that the drains are working well or are clogged. Therefore all design that follows is based on the assumption that there are no drains to be on the safe side.

Geotextile reinforcement was planned to be horizontally installed and distributed every 2 m on the face slope to the open river side as a protective measure against wave action, to reduce the waiting time between phases and to provide extra safety.

7.3 Stability analysis

In this section a detailed description of the design for the new alternatives is given. PLAXIS (strength reduction method) and to a lesser extent SLOPE/W (limit equilibrium method) were utilized to assess the problem.

7.3.1 Undrained analysis

A factor of safety of 1.3 was adopted here as minimum requirement to guarantee the safety during construction.

7.3.1.1 Phase 1

As illustrated in figure 7.2, the first phase goes up to level +4.0TAW, that means up to an embankment height of about 20 m. This phase was divided in 10 sublayers to be constructed in stages every 2 months approximately to allow for some consolidation and strength increase of the soft soil.

The following relationship (see also section §6.3.1 for the choice of parameters) was adopted to roughly estimate the undrained strength increase as consolidation proceeds:

$$\Delta c_u = 0.3 \Delta \sigma'_v \quad \text{with} \quad \Delta \sigma'_v = \Delta \sigma U \qquad (7.1)$$

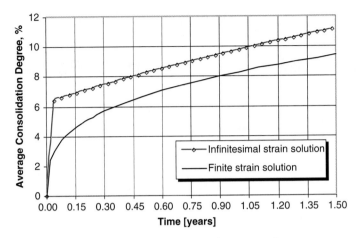

Figure 7.3. Consolidation degree vs. time

Table 7.1. Increase of undrained shear strength for phase 1

Sublayer	Level TAW	Time (year)	U (%)	$\Delta\sigma_v$ (kPa)	$c_u = c_{uo} + \Delta c_u$ (kPa)
1	−14.3	0	0	20	3.00
2	−12.3	0.16	4.7	40	3.28
3	−10.3	0.33	5.8	60	3.71
4	−8.3	0.50	6.6	80	4.20
5	−6.3	0.66	7.3	100	4.75
6	−4.3	0.83	7.8	120	5.35
7	−2.3	1.00	8.3	140	5.99
8	−0.3	1.16	8.7	160	6.66
9	+1.7	1.33	9.0	180	7.36
10	+4.0	1.5	9.4	220	8.09

where Δc_u is the undrained strength gain, $\Delta\sigma'_v$ is the effective stress increment, $\Delta\sigma_v$ is the total stress increment at the top of the layer, and U is the average consolidation degree at a time t.

As discussed in chapter 5, there are two procedures to predict the consolidation behavior, the classical theory (infinitesimal strain) and the finite strain theory (large strain). Both methods were attempted here based on test results and constitutive relations for consolidation given in chapter 6. Figure 7.3 illustrates for example the consolidation degree of the soft layer out of the two analysis based on excess pore water pressure dissipation. The results show that the large strain theory, which is a better approach to the actual behavior, gives a smaller mean consolidation degree than the classical theory. Then the use of the classical theory is underconservative in this case.

Table 7.1 summarizes all calculations of consolidation degree and undrained strength increase for each sublayer of phase one calculated with the large

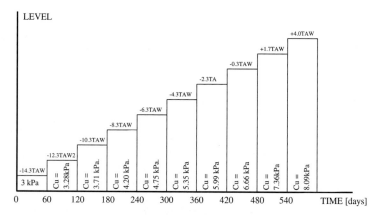

Figure 7.4. Construction program and available c_u

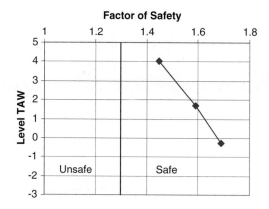

Figure 7.5. Factors of safety during construction of phase 1

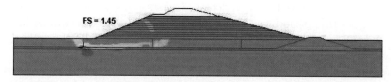

Figure 7.6. Undrained stability analysis of just terminated phase 1

strain consolidation theory. Moreover, figure 7.4 illustrates more clearly the construction program and the available c_u strength of the soft layer for each stage.

Figure 7.5 illustrates the results of the undrained stability analysis in terms of factors of safety of the embankment following the construction program showed in figure 7.4. Clearly, phase 1 can be built without concern about failure as a factor of safety of 1.45 is reached at the end of construction of phase 1 (Fig. 7.6).

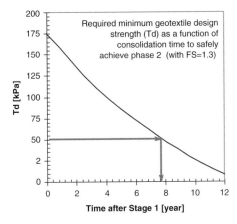

Figure 7.7. Geotextile design strength vs. waiting period between phase 1 and 2

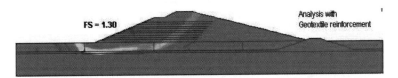

Figure 7.8. Undrained stability analysis of reinforced phase2

7.3.1.2 Phase 2

As far as phase 2 is concerned, the undrained stability analysis showed that the construction of phase 2 immediately after phase one would lead to instability. Then a long waiting period should be allowed until the soft material increases its strength up to a level enough to ensure safety.

Calculations showed that at least 13 years must be allowed. Given that such long waiting period was not practical for this project, an alternative solution had to be proposed, that was geotextile reinforcement in the embankment body.

As illustrated by figures 7.7 and 7.8, the geotextile reinforcement can indeed have a beneficial effect to reduce the waiting period between phases 1 and 2 provided that at installation they are long enough to anchor themselves in the sand and long enough to cross the potential slip surface.

For example, if a geotextile with a design strength of 50 kN/m is chosen, then the waiting time required is about 8 years instead of 13 years.

7.3.2 Drained stability analysis

The embankment was then analyzed under drained conditions to evaluate the stability in the long term. The results are illustrated in figure 7.9. As expected, the long term stability of the embankment is not critical even when

Figure 7.9. Drained stability analysis of finalized embankment without account of geotextile reinforcement

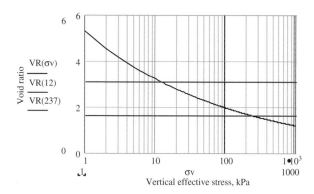

Figure 7.10. Void ratio (VR) vs. vertical effective stress

the geotextile reinforcement is not considered. If we consider geotextile reinforcement, the factor of safety would increase from 1.34 to 1.42.

7.4 Settlements

Two approximations of settlements were made. The first one makes use of the compressibility parameters (i.e. compression index) estimated from the constitutive properties (chapter 6) and the the second one, within the elasticity framework, makes use of a reasonable Young's modulus.

7.4.1 From Constitutive relationships

The maximum expected settlement was evaluated from the constitutive relationships of consolidation behavior of the soft material: equations 6.1 and 6.2. Figure 7.10 illustrates the the relationship e–σ'_v and the initial and final (after full construction) state of stress.

The maximum settlement is estimated with equation (7.2) where, H_0 is the initial thickness of the layer, C is the compression index, σ'_0 is the initial vertical effective stress and $\Delta\sigma$ is the total stress increment.

$$S_\infty = H_0 \frac{1}{C} \ln\left(\frac{\sigma'_0 + \Delta\sigma}{\sigma'_0}\right) \qquad (7.2)$$

88 Design of underwater embankment of soft soil

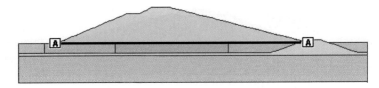

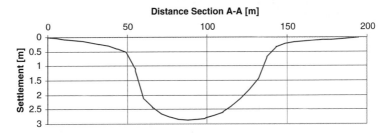

Figure 7.11. Settlement profile at section A-A (from PLAXIS)

Then, we get:
$$S_{\max} \approx 3.0\,\text{m}$$

7.4.2 From finite element program

The second method for settlement prediction is based on the theory of elasticity. The term "elastic settlement" is usually associated to "initial settlement" during undrained loading, however, elastic theory can also be used to estimate settlements due to undrained loading followed by consolidation. It will depend on the sort of parameters employed whether undrained or drained.

Figure 7.11 illustrates the total settlement, along section A-A, obtained from PLAXIS. We can see that the maximum settlement is about 3 m between the improved zones.

A good agreement is observed between the results coming from both methods which suggests that a reasonable estimation of the equivalent Young's modulus has been adopted.

7.5 Conclusions

It has been shown that although the starting value of undrained strength is quite low, it is still possible to carry out a safe construction splitting the construction in two main phases with a waiting period in between and introducing geotextile reinforcement. The reinforcement is required only to cope with the "rapid" construction of phase 1 but as soon as the foundation soil is allowed to consolidate and gain strength, the role of the geotextiles becomes a lot less important. In fact, stability analysis showed that the long term stability is assured even without geotextile reinforcement.

A maximum consolidation settlement of the order of 3 m can be expected between the improved zones. This settlement is expected to develop gradually in a long period of time. Judging from the consolidation predictions, a fast progress can be expected during the first couple of years. This predictions were confirmed later with measurements (see section 5.5).

Liquefaction risk is known to be an issue for hydraulically placed fills. However, given the low earthquake activity in Belgium, such risk was shown to be low. Safety factors against liquefaction were initially evaluated making use of expected values of cone pressure in the embankment sand and they were confirmed later with actual CPT data (see section 4.7).

8

Ground improvement by deep mixing

W.F. Van Impe & R.D. Verástegui Flores
Laboratory of Geotechnics, Ghent University, Belgium

P. Mengé & M. Van den Broeck
DEME, Belgium

8.1 Introduction

In recent years, an increased interest has been clearly demonstrated for design and construction on low bearing capacity soils, principally in low lands where many construction projects are conducted on soft alluvial clays, land reclaimed with dredged materials, highly organic soils, etc. In order to cope with such difficult subsoil conditions, various types of ground improvement methods have largely evolved in the last decades as described by Van Impe W.F. (1989, 1997).

The deep mixing (DM) method originated in the early 1970s in the Scandinavian countries and in Japan, almost simultaneously. The method could be classified as a permanent soil improvement technique with addition of cementing agents. Nowadays, binders such as cement, quicklime, fly ash, blast furnace slag, etc. are commonly employed to enhance mechanical and/or environmental properties of the natural soil. The cementing agents can be injected and mixed in place in either slurry or dry form. Furthermore, the mixing procedure can be purely mechanical or high-energy pressurized, both typically making use of rotating mixing tools.

The deep mixing method is currently of great interest and is often applied in near shore conditions for a number of applications such as improving bearing capacity, reducing settlements, reinforcing slopes, earthquake mitigation, etc. Worldwide statistics show the yearly growing importance of this technique (Rathmayer, 1996; CDIT, 2002).

In Belgium, deep mixing techniques are still not largely implemented. Nevertheless, scientific backed up experiences do shows the benefits that this ground improvement technique could bring in. In this chapter, local experiences on land and underwater (near shore) will be described. The on land experience makes use of dry deep mixing featuring mechanical mixing by means of a blade. The underwater near shore experience deals with the improvement of the underwater embankment foundation soil. It features wet deep pressurized intensive mixing.

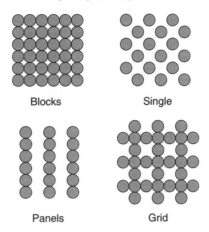

Figure 8.1. Types of application

8.2 Deep mixing applications

Worldwide DM methods have been implemented in special construction techniques for onshore and offshore civil works. They have also been used for improvement of foundations of a number of structures such as dams, embankments, tanks, bridges, retaining structures, high-rise buildings, etc. On land, the method has been employed on temporary stability works for deep excavations, protection of adjacent structures and stabilization of slopes.

Deep mixing can be executed in mass or column stabilization and both can be applied in many different ways. The soil can be stabilized either by forming columns of stabilized soil (so-called column stabilization) or by stabilizing the entire soil volume (so-called mass stabilization). Figure 8.1 gives some examples of the configuration of columns and figure 8.2 suggests the application for the combined mass and column stabilization.

Globally, local technologies are developed for new applications and for specific geographic areas often by innovative contractors who are seeking to develop their own variant of the method to match a particular project challenge.

The large number of existing DM techniques have been classified by Bruce (2001) on the following simple basis:

- Is the cementious material injected wet (W) or dry (D)?
- Is the binder mixed with the soil by means of rotary energy (R) only or is the mixing enhanced by high-pressure jet (J) methods?
- Is the mixing action only occurring near to the drilling tool (E), or is it continued along the shaft (S) for a significant distance above it, by way of augers and/or paddles?

Figure 8.3 shows the classification proposed on these basis. Four categories of methods have been identified: WRS, WRE, WJE and DRE.

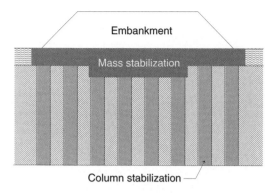

Figure 8.2. Combination of column and mass stabilization

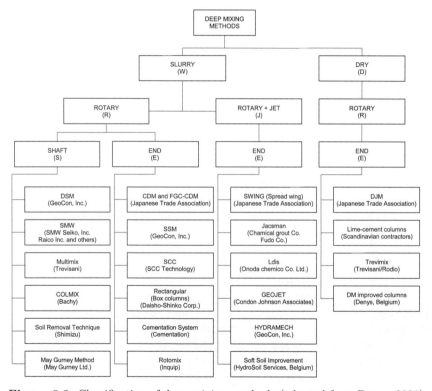

Figure 8.3. Classification of deep mixing methods (adapted from Bruce, 2001)

8.3 Mechanism of stabilization

The stabilizing agents are in most of the cases, Portland cement and lime, but also other binders have been more recently implemented. Some of these new binders have been designed for clayey soils with high natural water contents or

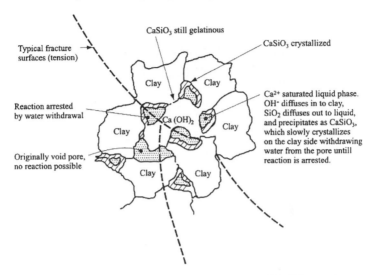

Figure 8.4. Lime stabilization mechanism (CDIT, 2002)

organic soils for which ordinary Portland cement or lime is not very effective. A brief description of the mechanism of stabilization of lime and cement is given in the following sections.

8.3.1 Stabilization with lime

When mixed with clayey soil, quicklime (CaO) reacts with pore water of the soil to become slaked lime ($CaO + H_2O \longrightarrow Ca(OH)_2$). This reaction takes places quickly and releases a large amount of heat. This brings in a reduction of the natural water content of the soil which already represent an improvement on its shear strength.

In presence of sufficient water the hydrated lime dissolves into Ca^{2+} and OH^-. Then, Ca^{2+} ions exchange with cations on the surface of the clay minerals. The cation exchange reaction alters the characteristics of the water films on the clay minerals. In general the plastic limit of the soil is increased, reducing the plasticity index. Furthermore, under a high concentration of hydroxyl ions (high pH), silica and/or aluminum in the clay minerals dissolve into the pore water and react with calcium to form a water insoluble gel of calcium-silicate or calcium-aluminate. This pozzolanic reaction goes on as long as the high pH condition is maintained and calcium exists in excess. Figure 8.4 illustrates the lime stabilization mechanism in which the product of the pozzolanic reaction cements the clay particles together.

8.3.2 Stabilization with cement-like binders

The most commonly used cement types for stabilization are Portland cement and Blast furnace cement. Portland cements are inorganic binders obtained

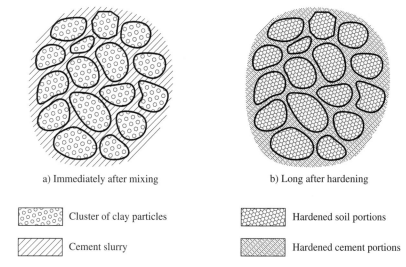

Figure 8.5. Cement stabilization mechanism (CDIT, 2002)

by grinding to a high fineness, Portland clinker alone, or most commonly in combination with calcium sulfate (gypsum) acting as a set regulator.

In ordinary Portland clinker, tricalcium silicate (C_3S) is the most abundant phase present in amounts between about 50% and 70%. Dicalcium silicate (C_2S) usually constitutes 15–30% of the clinker. Typical amounts of tricalcium aluminate (C_3A) are 5–10% and of the ferrite phase (C_4AF) 5–15%. During the hydration of the cement a C-S-H phase is formed and $Ca(OH)_2$ is released. The first hydration product has high strength which increases as it ages, while $Ca(OH)_2$ contributes to the pozzolanic reaction as in the case of lime stabilisation.

Figure 8.5 illustrates the cement stabilization mechanism. Immediately after mixing it is possible to identify clay clusters and cement paste as separate phases. Next, the strength of the stabilized soil will gradually increase due to pozzolanic reactions within the clay clusters and hardening of the cement paste.

Blast furnace cement is a mix of Portland cement and blast furnace slag and shows a similar stabilization mechanism. Finely powdered slag does not react with water but it has the potential to produce pozzolanic reaction products under high alkaline conditions. The SiO_2 and Al_2O_3 contained in the slag are actively released by the stimulus of the large quantities of Ca^{2+} and SO_4^{2-} from the cement, so that a hydration product is formed for which the long-term strength is enhanced.

The complicated mechanism of stabilization has been simplified by Saitoh et al. (1985) in figure 8.6 for the chemical reactions between clay, pore water, cement and slag.

96   Ground improvement by deep mixing

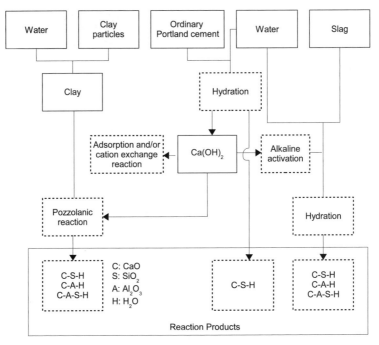

Figure 8.6. Chemical reaction among clay, cement, slag and water (Saitoh et al., 1985)

8.4 Methods of installation

The installation of columns stabilized with deep mixing methods requires the use of specially designed machinery which basically consists of:

- A binder feeding unit
- A soil mixing machine for injection of binder into the ground

Figure 8.7 illustrates a scheme of a typical DM column installation machinery. The binder feeding unit comprises various devices that measure the quantity of the ingredients in the admixture and transport it to the soil mixing machine. The plant generally include silos, automatic batching scales and a slurry or air pump.

A variety of soil mixing tools have been manufactured for various improvement purposes, ground conditions and special applications. In general, two categories of mixing tools can be identified:

Blade-based mixing tools: These tools have wide blades for excavation, in addition to paddles and/or short blades for cutting and mixing. The mixing process is mainly carried out at the tip (or close to the tip) of the tool (e.g. Fig. 8.8). The blades have a variety of shapes, dimensions and orientations.

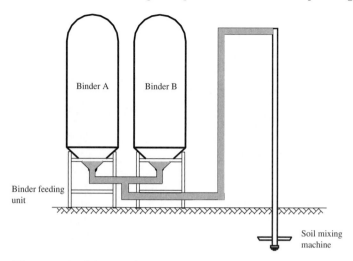

Figure 8.7. Scheme of equipment for DM column installation

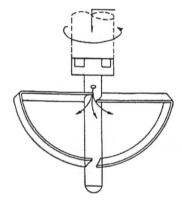

Figure 8.8. A typical mixing tool for the dry mixing method in Europe

Auger-based mixing tools: These tools have discontinuous or continuous helical augers for drilling, in addition to paddles and/or short blades for cutting and mixing. The mixing process is carried out in portions along the drilling shaft.

8.5 Belgian experience on on-land deep mixing

The aim of this section is to illustrate the Belgian experience with the dry deep mixing technique that could be implemented on, for example, improvement of an embankment foundation on land.

A broad research project devoted to investigate the performance of deep mixing methods for improvement of Flemish alluvial soils was carried out at Ghent University (Verástegui et al., 2005).

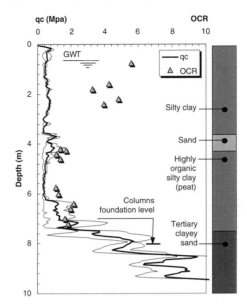

Figure 8.9. Natural soil profile at the test site

The research was focused on the improvement, on land, of soft clay and peat (often encountered in Flanders) with addition of cementing agents by means of dry mixing. Lime and different types of cement (i.e. Portland cement, blast furnace cement and others) have been employed. In the laboratory, the improvement of small cylindrical laboratory-mixed specimens has been followed up for a certain period of time. It has been found that blast furnace cements work quite well for both soil types. In the field, instrumented trial embankments built on improved and non-improved ground showed the benefit of different binder dosages when lime-cement columns are installed. The actual improvement in the testing site was assessed by field and laboratory testing.

8.5.1 Properties of untreated soils on land

The soil profile at the test site for the implementation of L-C column improvement was defined after an extensive field and laboratory testing campaign including piezocone penetration tests, vane tests, dilatometer tests and borings for sampling of disturbed and undisturbed specimens.

The CPT soundings showed the presence of soft alluvial soil in the upper 8 m of the profile overlying a clayey sand formation (Tertiary). Moreover, the soft layer was not homogeneous but it consisted of two main sublayers, corresponding to silty clay with sandy seams (from 0 to about 4 m below ground surface) and a highly organic silty clay (from 4 to about 8 m below the ground surface) with a sand content increasing with depth. Figure 8.9

Table 8.1. Physical properties of the natural soil

Index	Silty clay	Peat
Liquid limit	65.5	241.8
Plastic limit	22.8	135.0
Plasticity index	42.7	106.8
Natural water content	45.0	240.0
Organic content	1.7	18–30
Natural carbonates content	5.1	11.2
Sand fraction, %	28.3	29.0
Wet density, g/cm^3	1.7	1.2

shows a typical CPT profile and the OCR estimated by DMT. The highly organic silty clay is denominated "peat" for simplicity.

Disturbed samples taken at several depths from the silty clay and highly organic silty clay (peat) have been tested on physical properties. Some parameters of each soil type are summarized in table 8.1.

The undrained shear strength of the silty clay and the peat at the testing site has been determined by means of CPTU soundings, field vane tests, dilatometer tests and triaxial testing. Figure 8.10 summarizes all measurements. The undrained strength profile in the figure shows that c_u ranges, in general, from 20 to 40 kPa. The lowest values do obviously correspond to the peat.

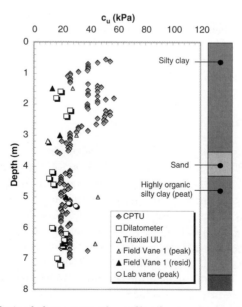

Figure 8.10. Undrained shear strength profile of natural soils at the testing site on land

8.5.2 Binders employed on land

In this research on dry mixing on land, quicklime and cement have been chosen as binders; however, different types of cement have been tried out (i.e. Portland cement, CEM I, composite cement, CEM II, and Blast furnace cement, CEM III). The name listing employed here complies with the standard EN 197-1; for example, CEM I 42.5 refers to a Portland cement with a nominal compressive strength of 42.5 MPa.

Lime and cement have been employed in different proportions (e.g. L/C 50/50, 20/80, 0/100, percentages in weight). Dry mixing was implemented in the laboratory as well.

The quantity of binder has been set to a range varying from 100 to 200 kg/m^3 (kg of binder per m^3 of natural soil). CEM I, CEM II and CEM III have been employed for the stabilization of silty clay. On the other hand, CEM II and CEM III have been used in case of the peat.

8.5.3 Lime-cement stabilization in the laboratory

8.5.3.1 Preparation of specimens

The natural soil samples collected from a number of borings have been first selected and then thoroughly homogenized prior to stabilization in the laboratory. A dough mixer has been employed in the laboratory for mixing of natural soil at the natural water content with the binders (added dry). A mixing time of about 5 minutes was implemented. Immediately afterward, small specimens ($H = 9$ cm, $\phi = 4.5$ cm) have been molded either by static compaction (for silty clay) or by pouring (for peat) into plastic split cylindrical molds.

The stabilized specimen have been sealed with paraffin and stored under water in a conditioned room at 20°C.

8.5.3.2 Unconfined compression tests

Unconfined compression (UC) tests have been carried out at specific time intervals up to 90 days after the preparation of the stabilized specimens (Verástegui et al., 2004). Some results have been summarized in figures 8.11 and 8.12. Figure 8.11 shows that the combination L/C-20/80 with blast furnace cement leads to the highest UC strength (UCS) for stabilized silty clay. In fact, a ratio $UCS_{stab}/UCS_{natural} \approx 40$ has been reached in 60 days with a dosage of 150 kg/m^3; moreover, UCS seems to still increase.

On the other hand, the combination L/C-20/80 with Portland cement shows little extra improvement after the first month; nevertheless, the ratio $UCS_{stab}/UCS_{natural}$ reaches a value of the order of 12. The composite binders with CEM II/B do show that, the higher the ratio of quicklime/cement the smaller the UC strength; however, quicklime plays a very important role on

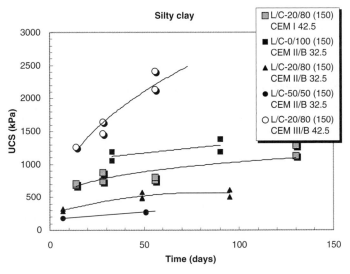

Figure 8.11. Unconfined compressive strength of silty clay

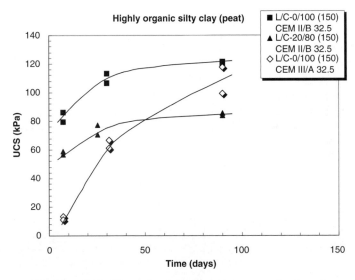

Figure 8.12. Unconfined compressive strength of peat

the quality of the mix, as the scatter of UCS decreases with an increasing amount of lime.

Figure 8.12 illustrates the development of the UC strength of stabilized peat with time. Clearly, the benefit of the lime here was less significant for the strength and mix quality. The UCS improvement on samples stabilized with CEM II/B seems to cease after 1 month, while specimens mixed with blast furnace cement, CEM III/A, show a slow but continuous increment. A ratio

$UCS_{stab}/UCS_{natural}$ ranging from 2 to 3 has been evaluated after 90 days. Note that the specimens have not been subjected to any overburden in the curing stage.

8.5.4 Lime-cement stabilization in-situ

To the extent of controlling the quality of the DM method itself in the field, a number of trial stabilized columns, $\phi = 0.6$ m, were installed on the on land test site with the dry mixing technique (more details were reported by Verástegui et al., 2004).

In the installation phase, the dry composite binder has been injected, by means of compressed air, at pressures not higher than 5 bar through a tubing down to the mixing tool. The DM column is formed below the mixing tool lifting the mixing auger while rotating continuously.

8.5.4.1 Quality control of lime-cement trial columns

Four trial columns were initially installed in the site on land. A scheme summarizing the characteristics and composition of each column is given in figure 8.13.

Column 1 has been stabilized with 85 kg/m^3 of unslaked lime, column 2 with 130 kg/m^3 of cement (CEM II/B-M 32.5), column 3 with 130 kg/m^3 of a blend (50/50) of unslaked lime and cement (CEM II/B-M 32.5), and column 4 with 170 kg/m^3 of unslaked lime.

Within the framework of quality control of stabilized columns, the extraction of the whole column would allow for a good evaluation of the homogeneity

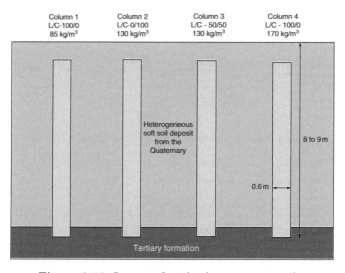

Figure 8.13. Layout of trial columns 1, 2, 3 and 4

Figure 8.14. Partial excavation of trial columns (a) No. 1 mixed with 85 kg/m^3 of L/C-100/0 (b) No. 2 mixed with 130 kg/m^3 of L/C-0/100 (c) No. 3 mixed with 130 kg/m^3 of L/C-50/50 and (d) No. 4 mixed with 170 kg/m^3 of L/C-100/0; all implementing CEM II/B-M 32.5

of the mix; however, such practice has proved to be expensive and difficult to put in practice. For that reason an alternative way has been chosen here; that is, partial excavation (to about 5 m below the ground level) and vertical borings through the entire column.

Columns 1 to 4, stabilized with different composite binders, have been excavated for visual inspection (Fig. 8.14). The inspection of the columns has

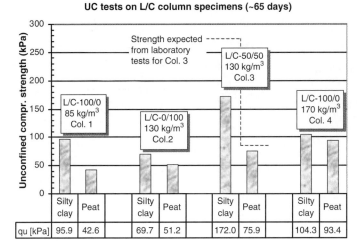

Figure 8.15. UC strength of specimens from the field inspection

taken place about 2 months after the installation. The following remarks were been made:

Column 1 (mixed with L/C-100/0, 85 kg/m^3) shows a rather uniform shaft diameter. Even though the soil in the neighborhood is very plastic, the visual inspection shows that the binder has been properly mixed.

Column 2 (mixed with L/C-0/100, 130 kg/m^3) shows some discrepancy on its diameter. It has also been noticed that the quality of the mix is not as good as in column 1 especially where the plasticity of the soil is high (little grains of hardened cement have been found there).

Column 3 (mixed with L/C-50/50, 130 kg/m^3) shows a very uniform, homogeneous and well shaped shaft. The quality of the mix seems to be quite good along the exposed portion of the column. The binder employed in this column is a blend of lime and cement.

Column 4 (mixed with L/C-100/0, 170 kg/m^3) shows a uniform shaft diameter as well. The binder seems to be very well mixed, as it was the case for column 1. The dosage employed here was rather high, still, no sign of binder spreading (outside the column) has been found.

Just after the inspection, a few specimens have been sampled (by means of horizontally pushed-in thin wall tubes) from the exposed section of the columns. Figure 8.15 summarizes the average results of UC tests on these specimens.

The results are consistent with the visual inspection. Column 2, in which the highest strength was expected, does not show a good performance. Indeed, the visual inspection had come across the fact that the mixing quality of Column 2 was the poorest. It has been suggested here that mechanical mixing of cement in a plastic soil is a very difficult task. However, when a blend lime-cement is employed, the lime reduces the plasticity of the soil facilitating the homogenization of the stabilized mass, which is later reflected in a higher strength.

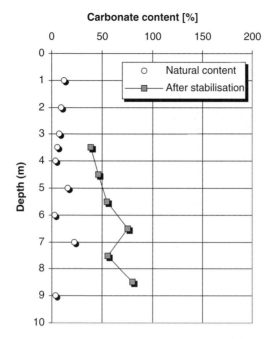

Figure 8.16. Distribution of binder along column 4, installed with the method implemented here, in terms of carbonate content

Moreover, figure 8.15 illustrates the strength expected from laboratory tests for column 3. It seems that the laboratory tests have over predicted the strength in the field with a ratio UCS_{lab}/UCS_{field} less than 2. Ratios ranging from 2 to 5 are usually reported in literature for dry mixing methods.

As there were no straightforward means of measuring the amount of cement in the soil due to its complex chemistry, it was decided to trace the amount of lime by means of a simple standard physical test. The test provides the amount of carbonates in the soil making use of a remotion agent (HCl). The results of the tests on specimens from the trial column 4 (that is mixed with lime only) are shown in figure 8.16.

Judging for the carbonate content of specimens from the column as compared to the natural lime content of the soil in that area, one could conclude that a rather uniform distribution of the binder has been achieved during the installation. This might also be the case for the other columns where composite binders were employed.

8.5.4.2 Trial embankments for testing the performance of improved soils

From the laboratory research outcome for the soils at the site on land, it was decided to use a combination quicklime/blast furnace cement (CEM III/B 42.5) at L/C-20/80 for silty clay and at L/C-0/100 for peat.

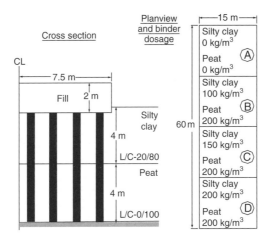

Figure 8.17. Layout of trial embankments

Next to the trial columns described in the previous section, 3 trial DM zones were built to evaluate the behavior of the foundation soil when subjected to an embankment load. Also a reference untreated foundation soil was loaded. The aim of these loading tests was to study the response of columns improved with different binder dosages.

Figure 8.17 illustrates the general layout of each embankment and the dosage per specific soil layer. Note that the dosage for the peaty layer was fixed to 200 kg/m^3 in all zones. The spacing (axis to axis) between columns was set to 1.8 m in a triangular arrangement. In each zone, the embankment fill aimed at a net surcharge of about 30 kPa.

Figure 8.18 illustrates the results of the assessment of column installation effects on the soil nearby a stabilized column within the column array for trial embankments. To that aim, dilatometer tests were performed in the close vicinity of a DM column before and 2 months after installation. It was clear that when comparing the state of the natural soil before and after installation in terms of the constrained modulus (correlated from DMT measurements) the installation effects seemed to be not detrimental at all. On the contrary, the modulus of the soil shows an increase, where more sandy soil is present. No stress relaxation was observed around the column due to the combined action of the mixing tool and the compressed air.

The in-situ evaluation of the improvement of this stabilized columns was carried out by CPT tests (5 months after installation). The CPT was performed through the axis of the column. Figure 8.19 illustrates the CPT profiles in the natural soil (untreated) and in the column axis. Clearly, a remarkable improvement, in terms of cone penetration pressure, can be observed in the upper silty clay layer where the ratio of $q_c column/q_c natural$ increases with depth to values of the order of 30 to 40. Similarly, an important improvement has been evaluated in the peaty layer with $q_c column/q_c natural$ ranging from 4 up to 7. The more sandy zones are clearly identified by the peaks of q_c.

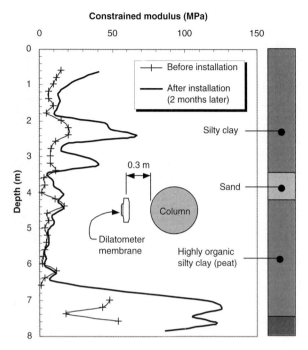

Figure 8.18. Evaluation of installation effects on adjacent soil within the trial embankment columns

Figure 8.20 shows the outcome of the settlement monitoring of the trial embankments by means of settlement tell tale plates. A period of about 2 months was set between the end of column installation and the initiation of trial embankments construction.

As expected, the reference embankment (A) shows the largest settlements and a very rational tendency was observed for the trial embankments on improved soil (B, C and D). The binder dosage for embankment B seems to be not high enough to allow for some significant benefit.

Also lateral deformations have been monitored by means of inclinometers installed on one side of the trial embankments (at about 1 m away from the side boundary). Figure 8.21 reflects these measurements, 1 month after loading. As expected, the horizontal deformations in a vertical close to embankment D are the smallest as compared to the values for embankment C. Embankment B, on the other hand, shows to induce by far larger lateral deformations, confirming the indication that such binder dosage employed in the soil deposit is insufficient to allow for a considerable improvement.

8.5.5 Remarks on the experience of dry deep mixing on land

The laboratory research for the application of deep mixing on land has shown good potential for the stabilization of silty clay and peat by blast furnace

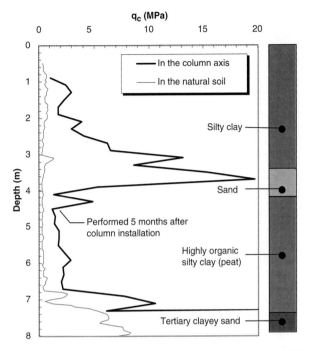

Figure 8.19. Evaluation of the improvement in situ by means of CPT on a column from embankment C

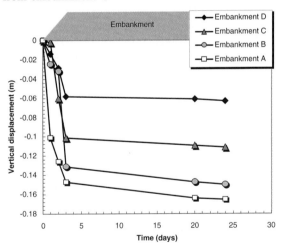

Figure 8.20. Settlement of the trial embankments

cements. A ratio $UCS_{stab}/UCS_{natural} \approx 40$ has been reached for silty clay (in 60 days) with L/C-20/80 (150 kg/m^3). The tests on peat indicate a slow but continuous improvement with a ratio $UCS_{stab}/UCS_{natural}$ ranging from 2 to 3, after 90 days.

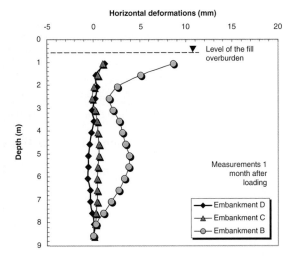

Figure 8.21. Settlement of the trial embankments

In the field, the assessment of the DM column installation effects (using this type of dry mixing method) with the column array for trial embankments allows to conclude for no stress relaxation. The evaluation of the improvement by means of CPT (performed 5 months after installation) shows a remarkable improvement in the silty clay layer where $q_c column/q_c natural$ increases with depth to values of the order of 30 to 40. In the peaty layer, on the other hand, $q_c column/q_c natural$ ranges from 4 up to 7. Moreover, from a quality control it was observed that the strength of specimens stabilized in the laboratory is greater than the strength achieved in the field by a factor less than 2.

The monitoring of trial embankments (on land) aimed at finding out the level of the benefit of the columns on the settlements and lateral deformations. A settlement reduction of about 65% was evaluated at the highest binder dosage implemented ($200 \, \text{kg/m}^3$). The lowest binder dosage of $100 \, \text{kg/m}^3$ was found insufficient to produce considerable improvement, at least on the soil conditions studied here.

8.6 Deep mixing assessment on the underwater site

The soil investigated here, as described in chapter 6, is a soft deposit of fine grained material, result of a prolonged sedimentation and self-weight consolidation process of dregs removed from the waterways within the harbor of Antwerp. In many harbor areas all around the world, there is an increasing need of reclaimed land. This fact has encouraged the design and ongoing construction in the Antwerp harbor of a partially submerged 27-m high sand embankment on the soft material previously mentioned. Obviously, the presence of such soft foundation layer caused concern for the overall stability; therefore partial improvement of the material by deep mixing was proposed.

This section focuses on the laboratory and field investigation carried out for the evaluation of the improvement of the soft fine grained soil with cement. Initially, the effect of several types of cement was studied in the laboratory. From those results a blast furnace cement was chosen as most suitable for the application in the field. Finally, a field inspection was carried out to asses the actual improvement of the deep mixing columns installed by the SSI technique (SSI is a technique patented by HSS, Dredging International-Belgium).

8.6.1 Properties of the artificially cemented soil in the laboratory

8.6.1.1 Preparation and mixing of specimen

The soil collected from the soft deposit was thoroughly homogenised and remolded prior to mixing with cementing agents.

A dough mixer was employed here to mix the soil and a slurry of cement. The dosage of binder for mixing with soil was set to 275 kg/m^3, the water/cement ratio of the slurry was set to 0.8 and a mixing time of about 10 minutes was implemented. This extended mixing time was meant to allow for more intensive mixing; however, only a slight difference in strength was observed when compared to specimens mixed for 5 minutes (less than 5% after 7 days).

Cylindrical specimen with a diameter of 57 mm and a height of 115 mm were prepared by pouring the mix into split plastic moulds. The moulds were later sealed with paraffin film and stored under water in a conditioned room at 10°C with no overburden whatsoever acting on the specimen. In addition, some specimen were cured under water at 20°C in order to study the effect of the temperature on the development of the improvement.

8.6.1.2 Binders

At the initial stage of this project, a number of different types of cement have been employed in the laboratory. A short description (according to EN 197-1) of the binders is given below:

- Binders A, B, and C are all blast furnace cements, CEM III. Binder C has the greatest blast furnace slag content (CEM III/B). Binders B and C classify at a nominal strength of 42.5 MPa while binder A has only 32.5 MPa.
- Binder D is a Portland cement, CEM I, with a nominal strength of 52.5 MPa.
- Binder E is a commercially available binder specifically designed for stabilization of soil.
- Binder F is a cement typically used for soil grouting purposes.

8.6.1.3 Compressive strength

A large number of unconfined compression tests have been performed at several time intervals (i.e. 7, 14, 28, 56, 84, 120, 240 and 550 days). The

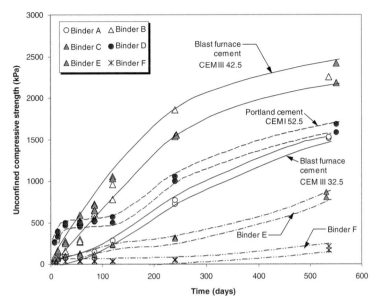

Figure 8.22. UCS of cement stabilised specimen cured underwater at 10°C

results of the testing programme on specimen cured under water at 10°C have already been reported by Van Impe et al. (2004) and are summarized in figure 8.22.

From the group of binders tested here, it seems that the blast furnace cements (binders A, B and C) perform quite well, showing a continuous increase of the UC strength. Binders B and C (both CEM III 42.5) do show an unconfined compressive strength of the order of UCS ≈ 2.2 MPa after 550 days. The Portland cement (binder D), on the other hand, allows for more rapid hardening in the first days. In fact, it shows the highest UC strength during the first month. However, the improvement provided by Portland cement seems to decline afterwards for some period to finally pick up again after some 3 months. The understanding of why systematically this "interval" of the interplay cement-soil occurs is subject to further research today. Anyhow, the final compressive strength of Portland cement remains lower than that given by the blast furnace cements B and C. The other binders (E and F) seem to produce little improvement for such high dosage (UCS < 0.7 MPa after 550 days).

The strain at failure of specimen cured under water at 10°C, illustrated in figure 8.23, was measured externally (from top to bottom cap of a triaxial cell) by LVDT. The figure provides some information about the ductility of the stabilised mass. In spite of some scatter it seems possible to establish a general tendency of behaviour for each binder mix. Overall, the strain at failure (ranging from 0.9% to 4%) decrease rapidly with increasing UC strength. The brittleness increases obviously with increasing UCS values. From the

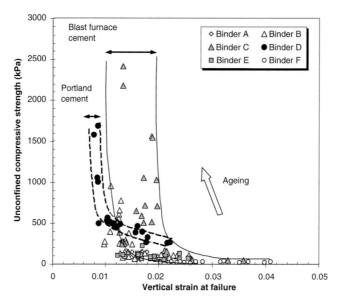

Figure 8.23. Strain at failure of cement stabilized specimen cured under water at 10°C

results it can also be deduced that, as ageing increases, specimen mixed with Portland cement tend to yield at smaller axial strains than specimen mixed with blast furnace cement, even tough the strength of specimen mixed with Portland cement is considerably lower.

8.6.1.4 Stiffness modulus

Measurements of small-strain modulus were also performed by means of bender element testing at different time intervals for some specimen (cured under water at 10°C) mixed with blast furnace cement (binder C) and with Portland cement (binder D) only.

The bender element test set up employed here is given in figure 8.24. The principle of this non-destructive method is simple and well know from literature (Dyvik and Madshus, 1985).

As an example, figure 8.25 illustrates the S-wave arrival time measured for specimen stabilised with blast furnace cement at several curing time intervals using an input sinusoidal pulse with a frequency of 4 kHz. Each specimen was tested for unconfined compression to measure UCS. As expected, a rather linear relationship between G_0, E_0 and UCS is observed.

Figure 8.26 summarizes the Young's modulus at small strain E_0 evaluated here for specimen mixed with Portland and blast furnace cement. The modulus for the Portland cement was found to be slightly higher but still, a single linear correlation has been proposed for both cements: $E_0 \approx 714 \cdot UCS$. Similarly, figure 8.26 illustrates the secant Young's modulus evaluated from

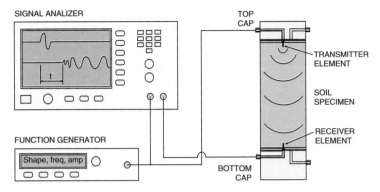

Figure 8.24. Bender elements testing setup

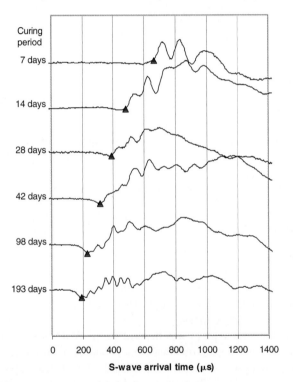

Figure 8.25. Shear wave arrival time measured at several curing time intervals on specimen mixed with blast furnace cement and cured under water at 10°C

unconfined compression tests. Even if trend shows some scatter, the data could be more or less linearly correlated to UCS as well. It has been estimated as $E_{s50} \approx 110 \cdot UCS$. This trend is considerably low when compared to the Japanese experiences reported by Saitoh et al. (1985) where

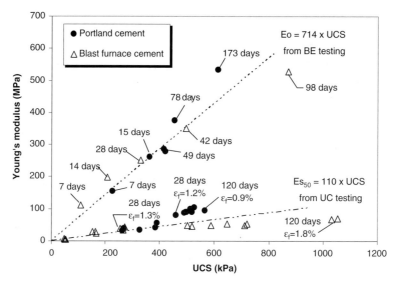

Figure 8.26. Young's modulus at small strain levels (E_0) and secant Young's modulus at 50% of deviatoric stress (E_{s50}) versus UCS

$350\,UCS < E_{s50} < 1000\,UCS$; however, it falls within the range of many other correlations proposed worldwide in the literature (Porbaha et al., 2000).

Overall, the modulus of the Portland cement is slightly higher than that given by the blast furnace cement. In general, E_0 remains about 7 times E_{s50}.

8.6.1.5 Effect of the temperature

In an attempt to more reliably recreate the conditions in the field, a large cylindrical specimen with a height $H \approx 0.8\,\mathrm{m}$ and diameter $\phi \approx 0.6\,\mathrm{m}$ was prepared in the laboratory employing blast furnace cement, with the aim of evaluating and monitoring the temperature changes due to exothermic reactions within the stabilized mass.

The virgin soil was kept at a temperature of 10°C prior to mixing. After mixing of the soil and blast furnace cement slurry in a concrete mixer, the stabilized mass was poured into a large plastic mold (also stored at 10°C and with the above mentioned dimensions) where eight temperature transducers (labeled T1, T2, ... , T8) were installed at different locations within the sample.

A few small cylindrical specimen were also prepared and cured (under water at 10°C) following the ordinary procedure described in a previous subsection.

The temperature measurements within the stabilized mass over a period of 56 days are illustrated in figure 8.27. The readings of all temperature transducers do show a common trend. Immediately after mixing a sudden temperature increase was observed. After 3 days a maximum temperature of about 25°C

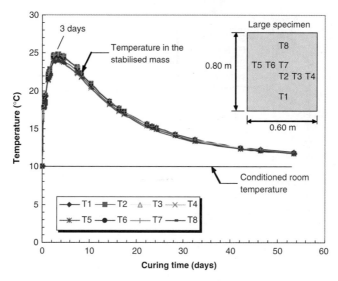

Figure 8.27. Hydration temperature monitoring on large specimen mixed with blast furnace cement and stored in a conditioned room at 10°C

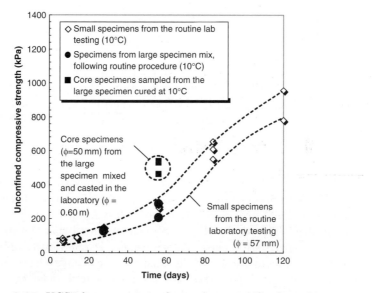

Figure 8.28. UCS of core specimen from a large stabilized specimen

was reached. Finally, the temperature in the large specimen seems to gradually decrease; after 56 days, the temperature (about 11.7°C) leveled out at values only slightly over the conditioned room temperature (10°C).

By the end of the temperature monitoring some core samples were taken from the large specimen. Figure 8.28 shows the UCS of such core samples. The

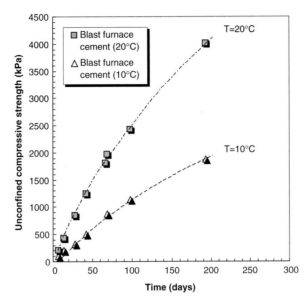

Figure 8.29. Effect of the curing temperature on the UCS of specimen mixed with blast furnace cement

figure also indicates the UCS of small specimen from the routine laboratory testing as described in a previous subsection. Clearly, the UCS of the large specimen cores doubles the UCS values of the small specimen. This suggested that the transient temperature increase due to the exothermic reactions within the large specimen were imposing such notable difference. Indeed, the larger the sample, the slower the heat dissipation and so the higher the UCS to be expected.

In order to study the effects of the curing temperature on the UC strength of the stabilized dredged material an extra series of tests has been carried out; this time on small specimen mixed with blast furnace cement, cured under water at 20°C. The results (Fig. 8.29) demonstrated that the strength of the samples stabilized with blast furnace cement is notoriously affected by the temperature. The hydration of the blast furnace cement clearly benefits from high temperatures; in fact, the UCS of samples cured under water (up to 200 days) at 20°C is, at all times, about 1.7 to 2 times larger than the UCS of specimen cured at 10°C.

8.6.2 Properties of the cemented soil in the field

The experimentation for the evaluation of properties of the cemented soil in the field consisted of core sampling of specimen from trial columns to proceed later on with unconfined compression tests in the laboratory.

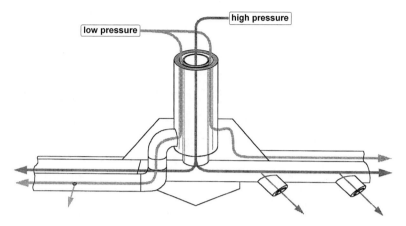

Figure 8.30. Mixing tool employed in the field

8.6.2.1 Installation of trial columns

The trial deep mixing columns ($\phi \approx 1.9$ m) were installed in the site (underwater) with the SSI technique from a jack-up platform.

Only blast furnace cement (binder C) was used for the field experimentation. The cement was mixed with water an transformed into a slurry ($w/c = 0.8$) on land. The cement slurry was pumped to the jack-up platform by means of floating pipes. In order to optimize the column installation rates the jack-up platform was provided with a moon pool to allow the installation of 22 to 24 columns in each zone covered by the platform. State of the art positioning systems ensured a very precise location of each column.

The SSI technique makes use of pressurized mixing by means of a mixing tool provided with 2 sets of nozzles distributed all along the full diameter of the column (Fig. 8.30). The mixing tool is fixed to a main drilling rod and each set of nozzles is connected to independent injection systems (Fig. 8.31). A high-pressure injection system (of the order of 20 to 30 MPa) cuts the soil and allows for intense mixing while the low-pressure injection system (up to 5 MPa) adds the remaining amount of cement slurry to fulfill the required dosage. Injection of the cement slurry takes place during the downwards and upwards operation of the drilling rod. A more detailed description of the installation and performance of the method can be found elsewhere (Van Mieghem et al., 2004). All drilling and injection execution parameters were automatically controlled to accomplish a binder dosage of 275 kg/m^3 approximately.

8.6.2.2 Evaluation of improvement in the field

A number of core specimen ($\phi = 100$ mm) obtained over the full length of the trial deep mixing columns were tested to evaluate the actual unconfined compressive strength.

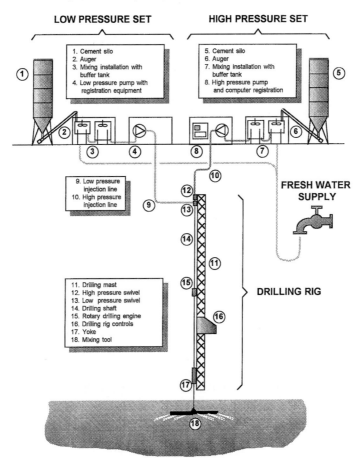

Figure 8.31. SSI set up and methodology

Figure 8.32 compares the strength evaluated from core specimen to that obtained from laboratory prepared specimen (after a curing period of 56 days). The UCS in the field ranges from 2 to 5 MPa in the upper 5 m and from 5 to 8 MPa in the lower zone, where a higher content of sand was observed. While, starting from the laboratory investigation, a quite optimistic UCS value remains below 0.9 MPa for similar conditions.

This discrepancy suggested that the ordinary practice of mechanical mixing (with a dough mixer) of specimen in the laboratory severely underestimates the actual strength of columns installed in the field with pressurized more intensive mixing procedures such as the SSI method (differences up to a factor of 2 to 5).

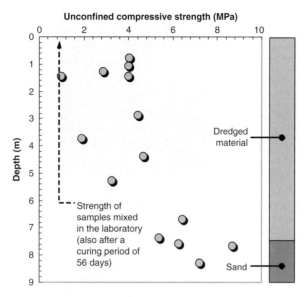

Figure 8.32. Unconfined compressive strength of core specimen from SSI-stabilized dredged material in the field

8.6.3 Laboratory versus in-situ behavior

In order to evaluate the reasons for discrepancy between laboratory and field test results, a laboratory reconstituted and mechanically mixed specimen on the one hand and undisturbed core samples from columns on site, on the other hand, were analyzed by means of Scanning Electron Microscopy (SEM), aiming at investigating the microstructure and composition of each specimen.

Figure 8.33 shows both specimen (4 cm × 4 cm × 1 cm) carefully cut with a water-cooled sawing system starting from stabilized samples, either mixed in the laboratory, or in the field by the SSI technique. It is obvious, already from this pictures, that the specimen differ in texture. Until the moment of the microscopic analysis, the sample from the laboratory was about 300 days old and had been kept sealed, under water, in a $T = 10°C$ conditioned room. On the other hand the sample from the field was approximately 270 days old; this specimen was cored from a trial SSI column about 3 months after its installation and then kept under water as well, until the day of SEM analysis.

The presence of large pores in a considerable amount in the laboratory specimen is evident. This is by far less pronounced in the SSI improved field specimen where a more compact and more homogeneous texture can be observed. At this point it may be stated that the mixing in the laboratory (by means of a dough mixer) could have caused the incorporation of air bubbles (large pores). Pores of smaller diameter observed in both specimen probably are produced during the cement hydration process.

Figure 8.33. Cement-stabilized specimen mixed by: (a) laboratory dough mixer, (b) pressurized in-situ mixing tool

Figure 8.34 does illustrate the same specimen but, this time, with an amplification factor of 1200. It is again quite clear that the micro structure is diverse. The specimen from the field has in general a much more homogeneous structure with a more regular distribution of hydration products, such as the calcium silicate hydrate (C-S-H phase) and the calcium hydroxide (CH). On the other hand, the mechanically mixed laboratory sample shows a rather heterogeneous skeleton where the unaged morphology of the C-S-H phase suggests still a lower degree of hydration.

Here it may also be suggested that the much more intensive high pressure mixing in the field did play an important role (the specific area around each soil particle could be reached by the binder, by far better). It seems that the high-pressure SSI mixing in the field has improved the distribution of cement around the soil particles and as a consequence a faster hydration and hardening has been taking place. In the laboratory, where purely simple

Figure 8.34. SEM analysis: (a) Laboratory specimen (b) field specimen at an amplification factor of 1200

mechanical mixing with a dough mixer was put into practice, the cement may have not been so well distributed and is only reaching clusters of soil particles.

Finally, figure 8.35 illustrates the samples from the laboratory and the field with an amplification factor of 1700. Also here the same pattern was observed; the structure of the field specimen looks much more homogeneous than the laboratory specimen. A matured C-S-H phase can be recognized in the field specimen together with uniformly distributed CH crystals that cover almost completely the soil. On the other hand, the mechanically mixed laboratory sample shows a much more heterogeneous composition including also ettringite (AFt phase) that is formed during the early hydration process (this phase is usually absent in matured and well hydrated cement pastes (Odler, 2000)). Overall, judging on the morphology of the different cement hydration products in the pictures, a by far less advanced degree of hydration could be perceived in the laboratory prepared samples.

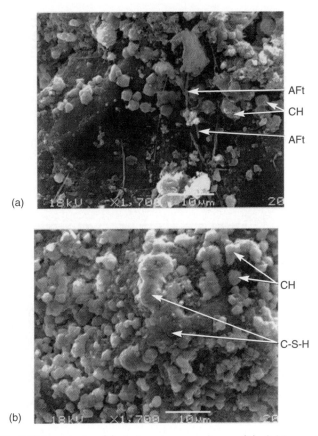

Figure 8.35. SEM analysis: (a) Laboratory specimen (b) field specimen at an amplification factor of 1700

9

Construction and monitoring of embankment

W.F. Van Impe & R.D. Verástegui Flores
Laboratory of Geotechnics, Ghent University, Belgium

J. Van Mieghem & A. Baertsoen
Ministry of Flanders, Belgium

P. Mengé & S. Vandycke
DEME, Belgium

9.1 Introduction

Today, the underwater embankment is still under construction. It has been built in stages by spreading sand in layers of approximately 2 meters allowing a period of time (1 to 2 months) in between. Today, a 70% of the total height of the embankment has been reached and now a much longer period is being allowed for consolidation.

Moreover, the quality of the embankment sand has been continuously controlled by means of CPT executed at several stages during the construction.

With the aim of increasing the safety, geotextile reinforcement was installed in the embankment slope at the open dock side.

Already before the initiation of the construction, instrumentation was mounted in the foundation layer to allow the monitoring of excess pore water pressures (PWP) and displacements due to the embankment load that could help to continuously check the behavior of the soil. Piezometers were installed at 3 different levels within the soft soil layer at different locations. Similarly, flexible tubes were placed at 4 locations across the dock to monitor vertical displacements by measuring water pressure changes (water height relative to a reference) with a probe that is displaced inside the tube.

The outcome of the monitoring of pore water pressures and settlement during the construction is given in the following sections. Also, the quality control of the embankment sand is briefly illustrated next.

9.2 Construction

The sand used for the underwater filling operation was mainly obtained from the excavation works and dredging residues of an almost simultaneous construction of a dock nearby. The sand was selected on the basis of its grain size distribution and content of fines so that optimum results of density and shear strength are obtained when it is hydraulically placed. Tests showed

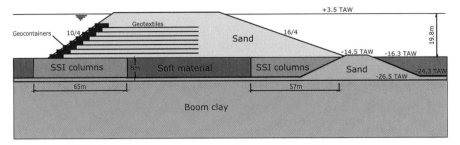

Figure 9.1. Layout of geotextiles and geocontainers

that the method of placing the sand could yield a sand shear angle $\phi > 32°$ and $c' \approx 0$, without any significant need of extra compaction after hydraulic placement.

The sand was transported from a temporary sand stock on land, by means of a sand pump, to a sand spreader vessel. The vessel was provided with a fall pipe with a 12-m wide horizontal spreader beam. With this system, adjusting the sand flow and the dynamic positioning of the vessel, sand layers with more or less uniform thicknesses could be placed.

After the installation of all Deep Mixing columns (already described in section 8.6.2), a first leveling sand layer with a thickness of about 1 m was placed over the whole area of construction (see Fig. 9.1).

The first embankment layer was built on top of the leveling layer. Also the first geotextile was placed and anchored on geocontainers as illustrated in figure 9.1. The geocontainers (approximate dimensions $3\,\text{m} \times 2\,\text{m} \times 30\,\text{m}$), filled with a mix of sand and and asphalt, were manufactured at the shoreline and the carefully placed by means of a floating crane provided with state of the art positioning systems (Fig. 9.2).

These operations were repeated for each 2 m sand layer to built up the embankment. Today, the embankment has successfully reached the water level. It is possible now, after about 2 years of construction, to see it emerging from the water (Fig. 9.3).

9.3 Quality control of the embankment sand

Quality control of the embankment sand was performed regularly at several stages during the construction by means of CPT tests. Out of cone penetration tests it was possible to observe the state of the hydraulically placed sand with depth. Moreover, some correlations of shear angle (ϕ) and relative density were attempted.

CPT tests were performed at several locations within the working area. Figures 9.4 and 9.5 show some examples of CPT results on sand overlying the SSI improved foundation soil. The figures show the quality control of the

Figure 9.2. Placing of geocontainers

Figure 9.3. Crown of the embankment beginning to emerge from the water (January, 2006)

embankment sand up to a level close to TAW 0.00 which is more or less the situation illustrated in figure 9.1.

Similarly, figures 3 and 4 show some examples of CPT results on sand overlying the non-improved (soft) foundation soil in between SSI improved soil areas.

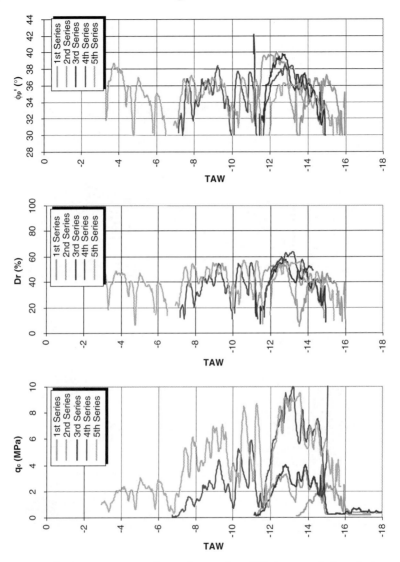

Figure 9.4. CPT test on sand overlying the SSI improved foundation soil, example 1

Overall, the cone penetration pressure q_c is observed to increase reaching values slightly greater than 10 MPa. However two different patterns can be identified.

Sand overlying the improved zone shows low q_c values at the interface with the foundation layer (TAW -16.00) and then it increases to reach maximum values at about 3 to 4 meters above such interface (close to TAW -13.00). These patterns are probably caused due to the arching effect taking place because of the presence of SSI treated columns. The arching effect causes the

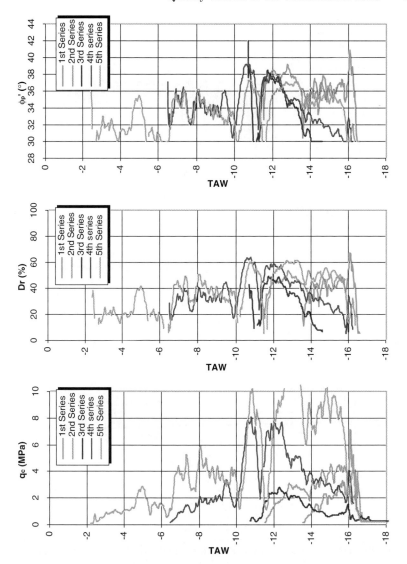

Figure 9.5. CPT test on sand overlying the SSI improved foundation soil, example 2

sand to be most stressed some distance above the interface with foundation layer while the sand below is barely receiving any surcharge.

On the other hand, sand overlying the non-improved foundation soil where there are no SSI columns show a more regular pattern of q_c with depth. In fact, an almost linear trend was observed.

The sand shear angle correlated from CPT complies, in all cases, with the design requirement of $\phi' = 32°$.

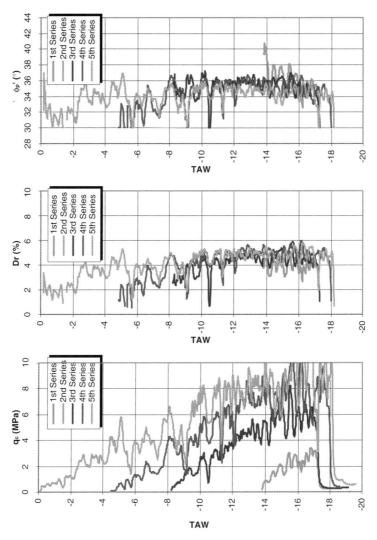

Figure 9.6. CPT test on sand overlying the non-improved foundation soil, example 1

9.4 Instrumentation and monitoring

Figure 9.8 illustrates the layout of the installed instrumentation to continuously follow up the excess pore water pressure and settlements during construction.

Piezometers were installed at various locations within the SSI-improved zones and the non-improved zones. They were also installed at different levels within the foundation layer.

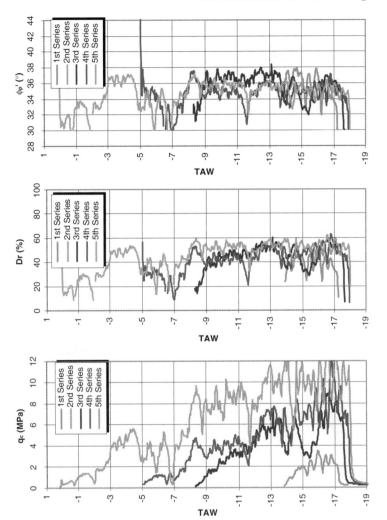

Figure 9.7. CPT test on sand overlying the non-improved foundation soil, example 2

Similarly, flexible tubes were installed at 4 locations across the dock (at the interface between embankment sand and foundation layer) to monitor vertical displacements by means of water head differences measured with a probe that is displaced inside each tube. Out of this method it was possible to evaluate settlement profiles within the SSI-improved zone and the non-improved zone.

9.4.1 Excess pore water pressure

Figure 9.9 summarizes the measurements of excess pore water pressure (PWP) in the foundation soil during construction. Significant differences can be

130   Construction and monitoring of embankment

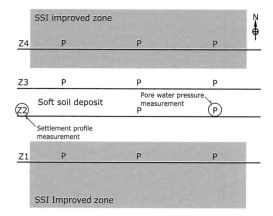

Figure 9.8. Plan view of the construction site showing a layout of the installed instrumentation

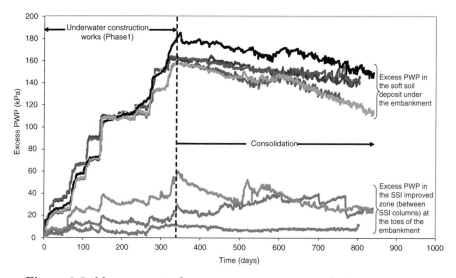

Figure 9.9. Measurements of excess pore water pressure during construction

observed between measurements within the SSI-improved zone and the non-improved zone.

As expected, the excess PWP in the improved zone is considerably smaller than in the non-improved zone, showing that the SSI columns are indeed taking up an important portion of the load. On the other hand, the excess PWP in the non-improved zone closely follows the stage construction loading history and it is only after about a year of construction that significant consolidation slowly takes place.

Instrumentation and monitoring 131

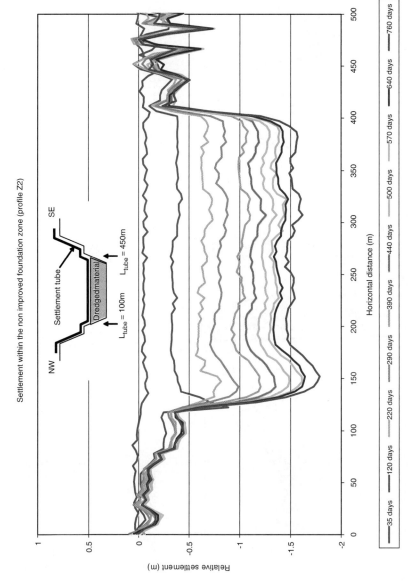

Figure 9.10. Settlement profile at the interface between embankment sand and foundation layer within the non-improved zone

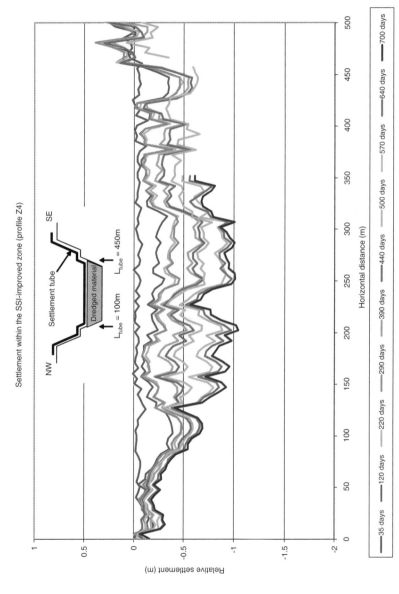

Figure 9.11. Settlement profile at the interface between embankment sand and foundation layer within the SSI improved zone

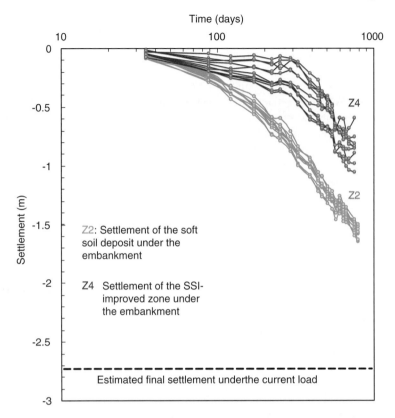

Figure 9.12. Settlement at the SSI-improved and nonimproved foundation zones

At this moment a more extended waiting period is being allowed so that enough strength is gained (due to consolidation) in the foundation soil to continue with the construction activities of the second phase.

9.4.2 Settlements

Figures 9.10 and 9.11 show the measured settlements profiles in the non-improved and the SSI improved foundation zones respectively. As expected, the largest settlements are observed in the non-improved area were up to now a maximum settlement of the order of 1.2 m was measured. On the other hand, the maximum measured settlement in the SSI-improved zone is of the order of 0.6 m.

Figure 9.12 compares the average settlements measured in both zones, SSI-improved and nonimproved. It can be observed that the current settlement in the nonimproved zone is about half the total settlement estimated for the present surcharge.

9.5 Conclusions

The construction of the sand embankment by hydraulic placement was shown to produce a sand mass with enough shear strength and cone penetration pressure to fulfil the design requirements.

Excess PWP and settlement measurements so far show values in agreement with the predictions at the design stage. In fact, the slow pore pressure dissipation coupled to the fast settlement development (as observed in the actual measurements) was already predicted with finite strain consolidation theories. This shows that when taking account of Key aspects of soil behavior, it is possible to properly model a complex problem such as this one and also to obtain acceptable predictions.

References

G. AAS, Stability of natural slopes in quick clays. Proc. 10th Int. Conf. Soil Mechanics and Foundation Engineering. Stockholm (1981).

J.H. ATKINSON, P.L. BRANSBY, The mechanics of soil: An introduction to critical state soil mechanics. McGraw-Hill, UK (1978).

L. BARDEN (1963) Stresses and displacements in a cross – anisotropic soil. Géotechnique, vol. 13, 198–210.

L. BARDEN (1965) Consolidation of clay with non-linear viscosity. Geotechnique, vol. 15, no. 4, 345–362.

R.A. BARRON (1948) Consolidation of fine grained soils by drain wells. Transaction, ASCE, vol. 113, 718–742.

S. BERNANDER, Active earth pressure built-up: A trigger mechanism in large landslides in sensitive clays. Technical report 49T, University of Lule (1981).

M.A. BIOT (1941) General theory of three dimensional consolidation. Journal of Applied Physics, vol. 12, no. 2, 155–164.

A.W. BISHOP, N. MORGENSTERN (1960) Stability coefficients for earth slopes. Geotechnique, vol. 10, 129–150.

L. BJERRUM (1967) Engineering geology of Norwegian normally-consolidated marine clays as related to settlements of buildings. Geotechnique, vol. 17, 81–118.

L. BJERRUM, Effect of rate of strain on undrained shear strength of soft clays. Proc. 7th Int. Conf. on Soil Mechanics and Foundation Engineering, Mexico (1969).

L. BJERRUM, Problems of soil mechanics and construction on soft clays, SOA report. Proc. 8th Int. Conf. on Soil Mechanics and Foundation Engineering, Moscow, URRS, vol. 3 (1973), 111–159.

A. BOUAZZA, W.F. VAN IMPE, W. HAEGEMAN (1996) Some mechanical properties of reconstituted Boom Clay. Jour. Geot. & Geol. Engineering, vol. 14.

B. BROMS, Progressive translationary landslides. Swedish Geotechnical Institute. Report No. 17 (1983).

D.A. BRUCE (2001) Practitioner's guide to the deep mixing method. Ground Improvement, vol. 5, no. 3, 95–100.

K.W. CARGILL (1984) Prediction of consolidation of very soft soil. Journal of Geotechnical Engineering, ASCE, vol. 110, no. 6, 775–795.

COASTAL DEVELOPMENT INSTITUTE OF TECHNOLOGY, The deep mixing method: principle, design and construction. A.A. Balkema, Rotterdam (2002).

E.H. DAVIES, G.P. RAYMOND (1965) A nonlinear theory of consolidation. Geotechnique, vol. 15, 165–173.

References

E. DE BEER, Shear strength characteristics of the Boom clay. Proc. of the Geotechnical Conf. on Shear strength properties of natural soils and rocks, Oslo, vol. 1 (1967).

P. DE WOLF, R. CARPENTIER, J. ALLAERT, J. DE ROUCK, Ground improvement for the construction of the new outer harbour at Zeebrugge – Belgium. Proc. 8th Int. Conf. on Soil Mechanics and Foundation Engineering, Helsinki, vol. 2 (1983), 827–832.

J.M. DUNCAN, State of the art: Static stability and deformation analysis. ASCE Geotechnical special publication No. 31, Stability and performance of slopes and embankments II, vol. 1 (1992), 222–266.

J.M. DUNCAN (1996) State of the art: Limit equilibrium and finite-element analysis of slopes. Jour. Geotechnical Engineering Division ASCE, vol. 122, no. 7, 577–596.

J.M. DUNCAN, S.G. WRIGHT, Soil strength and slope stability. John Wiley & Sons, USA (2005).

R. DYVIK, C. MADSHUS, Laboratory measurements of G_{max} using bender elements. Proccedings of the Conference on advances in the art of testing soils under cyclic conditions. Detroit (1985), 186–196.

P.J. FOX, J.D. BERLES (1997) CS2: A piecewise-linear model for large strain consolidation. Int. Jour. for Numerical Analysis in Geomechanics, vol. 21, no. 7, 453–475.

A. GABERC, Increase of subsoil bearing capacity beneath embankments. Proc. 13th Int. Conf. on Soil Mechanics and Foundation Engineering, New Delhi, India (1994), 759–762.

R.E. GIBSON, G.L. ENGLAND, M.J.L. HUSSEY (1967) Theory of one-dimensional consolidation of saturated clays. Geotechnique, vol. 17, no. 3, 261–273.

R.E. GIBSON, R.L. SCHIFFMAN, K.W. CARGILL (1981) Theory of one-dimensional consolidation of saturated clays, 2: Finite nonlinear consolidation of thick homogeneous layers. Canadian Geotechnical Journal, vol. 18, no. 2, s 280–293.

O. GREGERSEN, The quick clay landslides in Rissa, Norway. NGI publication, Oslo (1981).

W. HAEGEMAN, The measurement of small strain stiffness of terciary sands and clays with bender element tests in the triaxial apparatus. Proc. 13th Young Geotechnical Engineers Conf., Santorini, Greece (1999).

S. HANSBO (1979) Consolidation of clay by band-shaped prefabricated drains. Ground Engineering, vol. 12, no. 5, 16–25.

D.J. HENKEL (1960) The shear strength of saturated remoulded clays. ASCE Spec. Conf. on Shear Strength of Cohesive Soils, Boulder, 533–554.

K. ISHIHARA (1993) Liquefaction and flow failure during earthquakes. Geotechnique, vol. 43, no. 3, 351–415.

M. JAMIOLKOWSKI, C.C. LADD, J.T. GERMAINE, R. LANCELLOTTA, New developments in field and laboratory testing. Proc. 11th Int. Conf. on Soil Mechanics and Foundation Engineering, San Francisco, vol. 1 (1985), 57–153.

R.J. JARDINE, C.O. MENKITI, The undrained anisotropy of K0 consolidated sediments. Geotechnical engineering for transportation infrastructure, Barends et al. (eds.), Balkema, Rotterdam (1999).

D.C. KOUTSOFTAS, Undrained shear behaviour of a marine clay. Laboratory shear strength of soil, ASTM STP 740, Yong & Towsend (eds.) (1981), 254–276.

D.C. KOUTSOFTAS, C.C. LADD (1985) Design strengths for an offshore clay. Jour. Geotechnical Engineering Division ASCE, vol. 111, no. 3, 337–355.
R. LANCELLOTTA, Geotechnical engineering. AA Balkema, Rotterdam, Netherlands (1995).
C.C. LADD (1991) 22nd Karl Terzaghi Lecture: Stability evaluation during staged construction. Jour. Geotechnical Engineering Division ASCE, vol. 117, no. 4, 540–615.
C.C. LADD, R. FOOT (1974) New design procedure for stability of soft clays. Jour. Geotechnical Engineering Division ASCE, vol. 100, no. GT7, 763–785.
Z. LECHOWICZ, An evaluation of the increase in shear strength of soft soils. Advances in understanding and modelling the mechanical behaviour of peat, den Haan, Termaat & Edil (eds.), Balkema, Rotterdam (1994), 167–178.
S. LEROUEIL, J.P. MAGNAN, F. TAVENAS, Embankments on soft clays. Ellis Horwood Limited, England (1990).
T. MATSUI, KA-CHING, Availability of shear strength reduction technique. ASCE Geotechnical special publication No. 31, Stability and performance of slopes and embankments II, vol. 1 (1992), 445–460.
G. MESRI ABDEL-GHAFFAR (1993) Cohesion intercept in effective stress stability analysis. Jour. Geotechnical Engineering Division ASCE, vol. 119, no. 8, 1229–1249.
J.K. MITCHELL, R.G. CAMPANELLA, A. SINGH (1968) Soil creep as a rate process. Journal of the Soil Mechanics and Foundations Division, ASCE, vol. 94, no. SM1, 231–253.
J. NUYENS et al., National Report 10 – CPT in Belgium in 1995. Proc. Int. Symposium on Cone Penetration Test, Sweden, vol. 1 (1995), 17–27.
I. ODLER, Special inorganic cements. E&FN Spon, London (2000).
A. PORBAHA, S. SHIBUYA, T. KISHIDA (2000) State of the art in deep mixing technology. Part III: Geomaterial characterization. Ground Improvement, vol. 4, no. 3, 91–110.
I.C. PYRAH (1996) One-dimensional consolidation of layered soils. Géotechnique, vol. 46, no. 3, 555–560.
H. RATHMAYER (1996) Deep mixing method for soft subsoil improvement in the Nordic countries, Grouting and Deep Mixing. Proc. 2nd Int. Conf. on Ground Improvement Geosystems, Balkema, 2, 869–877.
P.K. ROBERTSON, C.E. WRIDE (1998) Evaluating cyclic liquefaction potential using cone penetration test. Canadian Geotechnical Journal, vol. 35, no. 3, 442–459.
P.W. ROSCOE, A.N. SCHOFIELD, C.P. WROTH (1958) On the yielding of soils. Geotechnique, vol. 1, 22–52.
S. SAITOH, Y. SUZUKI, K. SHIRAI (1985) Hardening of soil improved by deep mixing method. Proc. Int. Conf. on Soil Mechanics and Foundation Engineering, 11, San Francisco, Aug. 1985. vol. 3.
R.L. SCHIFFMAN (1980) Finite and infinitesimal strain consolidation. Journal of the Geotechnical Engineering Division, ASCE, vol. 106, no. GT2, s 203–207.
R.L. SCHIFFMAN, Theories of consolidation: a comparative study. Proc. Symposium on Developments in Geotechnical Engineering, Bangkok, Thailand, January (1994).
R.L. SCHIFFMAN, S.K. ARYA, One-dimensional consolidation. Numerical methods in geotechnical engineering, Desai & Christian (eds.), McGraw-Hill (1977), 364–398.

References

A. SCHOFIELD, P. WROTH, Critical state soil mechanics. McGraw-Hill, UK (1968).

H.B. SEED, I.M. IDRISS (1971) Simplified procedure for evaluating soil liquefaction potential. Journal of the Soil Mechanics and Foundations Division, ASCE, vol. 97, no. SM9, 1249–1273.

F. TAVENAS, Some aspects of clay behaviour and their consequences on modelling techniques. Laboratory shear strength of soil, ASTM STP 740; Yong, Towsend (eds.) (1981), 667–677.

F. TAVENAS, S. LEROUEIL (1987) State of the art on laboratory and in-situ stress-strain-time behaviour of soft clays. Proceedings of International symposium on geotechnical engineering of soft soils, Ciudad de Mexico, Agosto 1987, vol. 2.

D.W. TAYLOR (1937) Stability of earth slopes. Journal of Boston Society of Civil Engineering, vol. 24, 197–246.

K. TERZAGHI, Erdbaumechaniek auf bodenphysikaliser Grundlage. Leipzig Deuticke (1925).

USACE (1990) Engineering and Design – Settlement Analysis. Publication Number: EM 1110-1-1904.

P. VAN IMPE, Consolidatie van verzadigde, sterk samendrukbare poreuze media. Final degree work, Faculty of applied sciences, University of Ghent (1999).

W.F. VAN IMPE, Foundation engineering problems at the construction of the new Zebrugge harbour. Internal report, Laboratory of soil mechanics, University of Ghent (1985).

W.F. VAN IMPE, Evaluation of deformation and bearing capacity parameters of foundations from static CPT results. Proc. 4th Int. Geot. Seminar, Singapore (1986).

W.F. VAN IMPE, Soil improvement techniques and their evolution. AA Balkema, Rotterdam, Netherlands (1989).

W.F. VAN IMPE (1997) Soil improvement experiences in Belgium: part III. Case histories. Ground Improvement, vol. 1, no. 4, 179–191.

P.O. VAN IMPE (1999) Consolidation of very compressible saturated porous media. Final degree work, Faculty of Applied Sciences, Ghent University. In Dutch.

W.F. VAN IMPE, Deep mixing. Internal report, Laboratory of soil mechanics, University of Ghent (2000).

W.F. VAN IMPE, E. DE BEER (1984) Contribution to the analysis of landslides in quick clay. International symposium on landslides, 4, Toronto, Sept. 1984. Vol. 2.

W.F. VAN IMPE, R.D. VERÁSTEGUI FLORES, M. VAN DEN BROECK, P. MENGÉ (2004) Stabilisation of dredged sludge with different types of binders. Proc. Int. Symp. on Engineering Practice and Performance of Soft Deposits, Osaka, 197–200.

R.D. VERÁSTEGUI FLORES (2001) Geotechnics of staged construction. Final degree work, Faculty of Applied Sciences, Ghent University.

R.D. VERASTEGUI FLORES et al., Lime-cement mixing for improvement of alluvial soft Belgian soil. Proc. Int. Symp. on Engineering Practice and Performance of Soft Deposits, IS-OSAKA (2004), Japan 241–246.

R.D. VERASTEGUI FLORES, W.F. VAN IMPE, P. AFSCHRIFT, W. CROMHEEKE (2005) Lime cement columns in alluvial soft soil. Proc. Int. Conf. Soil Mechanics and Geotechnical Engineering, 16, Osaka, vol. 3.

Subject Index

behavior of clay 3–5, 11–16
bender element test 112–13
cone penetration test 50, 52–6, 72, 98, 106, 124, 126–9
consolidation analysis 7, 29, 62, 64
constant rate of strain consolidation test 77–8
critical state 14–15, 79
deep mixing 2, 77, 81, 91–122
dilatometer test 29, 98–9, 106–7
finite strain theory 57, 64–7, 69–70, 77, 84, 134
geotextile reinforcement 81, 83, 86, 88, 123–4
ground improvement 2, 91–122
hydration of cement 94–5
hydration of lime 94
infinitesimal strain theory 8, 57, 58–64, 66, 69–70, 84
k_0-consolidated clay 12, 15, 20–3
limit equilibrium method 25, 30, 83
liquefaction assessment 49–50
normalized behaviour 13–14, 18, 76
normally consolidated clay 4–6, 11, 18
on-land deep mixing 97–109
overconsolidated clay 4, 14–15, 72, 78, 83

scanning electron microscopy 73–4, 75, 119, 121–2
sensitive clays 27, 35–49
slope failure patterns, 25–6
soft soil improvement (SSI) 2, 52–3, 110, 117–19, 125–133
soft soils 4–8
staged construction 5, 8, 20, 28, 67, 70, 85
stiffness of treated soil 107, 112–14
strength reduction method 25, 31–3, 35, 51, 83
stress induced anisotropy 20
structural anisotropy 20
temperature effect on hydration 114–16
unconfined compressive strength 100–1, 104, 111, 115–16, 119
underwater deep mixing 109–22
underwater embankment 4, 67, 81–9, 123–5
undrained shear strength 7, 11–24, 29, 76, 83, 88, 99
undrained strength anisotropy 20–2
undrained strength ratio 13–14, 19, 22, 76